Atmospheric Processes and Human Influence

Paul Warburton

Head of Geography, Manchester High School for Girls

Collins Educational

An imprint of H...

Contents

Skills matrix
(distribution of numbered tasks)

Chapter	Understanding: text/newspapers/ classification	Graphical mapping methods/ annotated diagrams	Analysis of data: tables/ graphs/diagrams	Analysis of photos	Analysis of maps	Statistical analysis/ methods/ calculations	Values enquiry	Project work/library research/ fieldwork	Writing: essays etc.
1	3, 5, 6, 10, 15	2, 13	4, 5, 12, 14, 15	8, 9, 16	1, 16			7	11
2	4, 6, 11, 13, 19	3, 17	2, 4, 6, 7, 8, 9, 10, 11, 15, 16, 17, 19	15, 16	1, 2, 12, 13, 14, 15, 20, 21, 22	3, 5, 18			23
3	10	15, 16, 17	1, 2, 3, 4, 8, 11	4, 9	5, 6, 7, 16, 18			12, 17	13, 14, 19
4	1, 3, 4, 5, 12, 22, 24, 25, 26, 28 29, 30	10, 15	2, 6, 7, 8, 9, 11, 14, 20		13, 14, 16, 18, 20, 21, 23, 24		4, 5, 12, 26, 27, 29	17	31, 32
5	1, 8, 9, 11, 15, 17	10, 12	3, 4, 13	2, 11, 14	2		1, 8, 16, 19	6, 7, 16	5, 14, 18, 19
6	7, 12, 14, 16, 21	1, 11, 21	4, 7, 8, 16	3, 6, 16, 18	2, 4, 5, 10, 12, 13, 15, 16, 19				9, 17, 20
7	3, 14, 15, 19, 21, 24, 26	7, 10, 13, 27	3, 4, 5, 8, 9, 11, 12, 18, 23	1, 8, 21	2, 6, 7, 8	20			16, 17, 22, 28
8	3, 5, 7, 9, 13, 22, 23, 25, 26		2, 8, 10, 12, 19, 24, 25, 26	15	4, 5, 6, 12, 14, 15, 16, 17, 18, 22		24	11	5, 27
9	2, 3, 7, 14, 15, 20		3, 4, 7, 9	2, 6, 10, 16, 18	1, 9, 10, 11, 12, 13, 17, 18, 19				5, 21
10	1, 4, 12, 13, 15, 17	3, 7, 10, 11, 14	1, 3, 4, 5, 11, 12, 16	2	3, 8, 9, 12, 13		16	13	6
11	3, 6, 7, 11, 14, 15	1, 5, 15	2, 4, 9	1, 12, 14	2, 3, 14	7	9, 10		8, 16
12	1, 2, 12, 16, 17, 23, 24	14, 18, 20, 22, 25, 26	3, 4, 8, 15, 19, 21, 26, 27	4, 13	3, 5, 6, 7, 9, 10, 11, 12, 26	20		25	17
13	3, 4, 5, 9, 14, 16, 18, 19	6, 13	1, 2, 7, 8, 9, 11, 12	10	6, 7, 8, 11, 15				17, 20

To the student

The majority of people rarely give any thought to the atmosphere. Like soil and water, we accept that it exists and assume it will always provide us with the oxygen that we need to breathe. This casual attitude has also led many to believe that the atmosphere is a huge rubbish tip, like the oceans, where we can endlessly dump our waste.

Although the workings of the atmosphere have fascinated some individuals, most of us find it a difficult subject. Many aspects involve some advanced chemistry and physics. In fact, it has been my experience that many teachers and students avoid studying weather and climate on GCSE and A-level courses, or only give it superficial coverage. This is a pity, considering the many links between meteorology and other areas of geography.

It was my intention to try to remedy this situation by writing a textbook which avoids jargon and uses very clear language to overcome these problems. Throughout the book I have stressed links between human behaviour and atmospheric processes. This not only establishes relevance but is also realistic, as there are many two-way interrelationships between people and the atmosphere including hazards, agriculture and energy use. The media have shown this in recent years by giving increasing attention to issues like global warming, ozone depletion and smog. You should also become aware of how atmospheric processes interact with other parts of the physical environment: hydrology, oceans and seas and vegetation.

This book has been written with A-level geography students in mind, although it should also appeal to a wider range of students without a background in geography. You will need to make full use of the diagrams, photographs and maps to help you to understand the text. There are also many different activities including analysis of satellite images and weather maps, statistical exercises, interpretation of text and essays. These will test a range of skills relevant at A-level and above.

Further aids to your studies include plans at the beginning of each chapter stressing the main themes. Explanations of key terms then occur either in the text or glossary (shown in bold), while theories and individual case studies are distinctly separated from the main text. Each chapter concludes with a summary to help you with revision. Although many professional meteorologists require a training in physics, I have stressed the importance of a geographical approach. There are many references to the contribution geographers have made to the understanding of the atmosphere, and the application of their skills is supplied in a number of contexts.

If we accept that one of the objectives of geography is to enhance environmental awareness and sensitivity, I hope this book will also increase your awareness of the drama and beauty of the atmosphere. Many of us have been impressed by a stunning sunset or the power of a hurricane; atmospheric phenomena do not always have to be studied in an academic way. It will only be when we learn to care for the atmosphere and understand how it works that we will want to protect it for our future.

Paul Warburton

1 The dynamic atmosphere

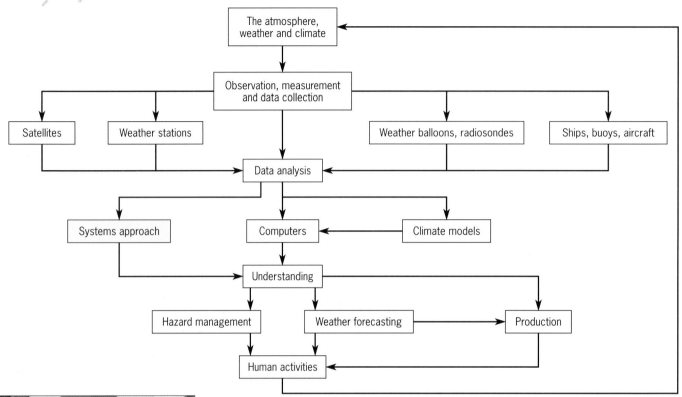

Figure 1.1 Umbrellas are out in force as rain stops play, Wimbledon Tennis Championships, England, 1991

1.1 Introduction

The atmosphere is constantly changing. We have all seen dramatic news reports of such events as a **hurricane** in the Caribbean or the **monsoon** in South-East Asia. You may have watched one of the satellite sequences on a television **weather** forecast and observed swirls of cloud developing and crossing the British Isles. Under those clouds perhaps light rain and winds develop into a powerful storm that eventually passes and fine weather returns.

Weather and climate

The weather is probably the most widely discussed aspect of the physical environment. Good weather can bring a surge of activity including outdoor sports and summer holidays. In Britain, though, it never seems to be quite right! Crops may wither in the heat and gardeners complain about hosepipe bans. We are all likely to have had plans upset by a sudden change in the weather (Fig. 1.1) or complained because the **climate** in our part of the country is too wet or dry. The Manchester Weather Centre copes with about 120 000 enquiries a year from people wanting to know whether the sun will shine for their wedding or garden party, or if there will be more frosts if they put out their bedding plants.

We often use the words weather and climate very loosely, though, without being aware of their precise meaning. Weather (Fig. 1.2) refers to the state of the atmosphere over an area for a short period of time. We tend to use the **elements** of the atmosphere, such as rainfall, **pressure** and wind, to describe the weather. Both the temporal (relating to time) and spatial (relating to space) scales involved are

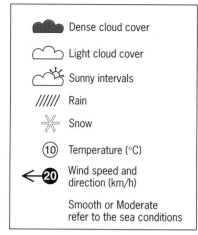

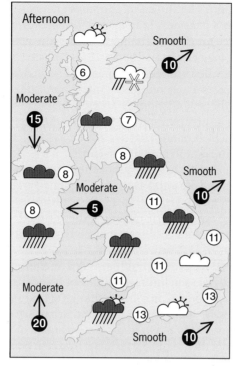

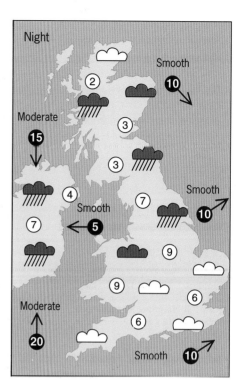

Figure 1.2 Part of a weather forecast (*Source: The Guardian*, 21 April 1994)

Table 1.1 Manchester, Ringway Station: climate statistics (*Source:* Met. Office)

Months	January	February	March	April	May	June	July	August	September	October	November	December
Temperature (°C)												
Average max.	6.4	6.6	8.9	11.6	15.3	18.2	19.6	19.5	17.0	13.7	9.1	7.1
Average min.	1.3	1.2	2.5	4.3	7.3	10.2	12.0	11.9	10.0	7.5	3.6	2.0
Precipitation (mm)	69.0	50.0	61.0	51.0	61.0	67.0	65.0	79.0	74.0	77.0	78.0	78.0

smaller than those for climate. Climate (Table 1.1) can be simply defined as the average weather over an area. Although we refer to larger temporal and spatial scales, we describe the climate of an area using the same elements. A weather balloon, for example, might record temperatures and **humidity** at a very specific location for a particular moment in time. In contrast, climate data for the British Isles covers a much larger area and is usually based on a period of 30 to 40 years.

The atmosphere and human activity

We need to understand something of the chemistry and physics of the atmosphere to be able to make sense of climatic and weather patterns. For example, we cannot fully understand the damage being caused to the **ozone layer** without knowing how ozone is formed and what gases and chemical reactions cause its destruction. The scientific study of the atmosphere is called **meteorology**. As our knowledge of atmospheric processes grows, we will also be able to recognise and understand the harmful effects that many human activities are having on the atmosphere.

This book emphasises the two-way links between the atmosphere and human activity. Chapter 1 encourages you to think further about the ways in which climate and weather affect human behaviour. It also begins to show how our activities are increasingly damaging the workings of the atmosphere and creating climatic hazards like **global warming**. In addition, we will see how vital information is gathered about atmospheric processes which helps us understand the atmosphere and make accurate weather forecasts and projections of climate change.

?

1 Use Figure 1.2 to describe the weather over the British Isles and how it will change towards the end of the day.

2a Complete a climate graph based on the data in Table 1.1.
b Use your completed graph to describe the climate of Manchester. You could make comparisons with the area where you live.

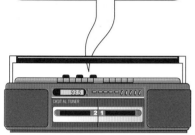

The sequence of depressions that has brought so much rain and cold weather over the last few weeks is coming to an end. A high pressure system to the south-west will develop over the next few days bringing warmer conditions. It seems that summer has finally arrived!

Figure 1.3 Weather forecast: warm weather on the way

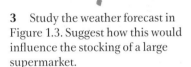

3 Study the weather forecast in Figure 1.3. Suggest how this would influence the stocking of a large supermarket.

Figure 1.4 Well-formed savoy cabbage ruined by heavy frost and snowfall, February, England

Chapter 2 follows the movements of energy through the atmosphere and looks at the temporal and spatial variations in the solar energy received by the earth. The implications for the use of the sun's energy as a power source are also examined. Chapter 3 considers long-term and short-term climate change. It examines the evidence for climate change as well as the natural and human-made causes. This follows logically from Chapter 2, as climate change has much to do with disruption to energy flows in the atmosphere.

Chapters 4 to 8 emphasise major atmospheric processes and their links with human behaviour. Both Chapter 4, on global warming, and Chapter 5, on the ozone layer, consider how energy flows and the composition of the atmosphere are being affected by human activities as well as physical processes, and how these changes are seriously affecting life on earth.

Chapter 6 focuses on atmospheric movement, describing and accounting for average and seasonal variations in global circulation. The consequences of air movements for vegetation and agriculture are also considered. Chapters 7 and 8 examine atmospheric moisture, the various forms that this can take and how they develop, and the problems caused by extremes of **precipitation**. We will also look at the issue of **acid rain**.

The climates of the mid-latitudes, including **depressions** and **anticyclones**, and tropical areas, including the monsoon and hurricanes, form the subject of Chapters 9 to 11. Chapter 12 involves a change of scale and examines a variety of local climates, including mountains and urban areas.

Chapter 13 is a case study of the climate of Brazil. The chapter is designed to bring together many of the ideas from the book in a regional context. There will also be a number of activities designed to test your understanding of many of the concepts that you will have studied.

1.2 The importance of weather and climate

Although our lives in Britain have become very sophisticated, we can still be affected by the weather and climate. Many people in our society need up-to-date information about the weather for their jobs. A number of companies also benefit from advance notice of changes in the weather. For example, Sainsbury's, the supermarket chain, uses meteorological information to time when to stock or de-stock items like soft drinks and ice cream. A garage stocking anti-freeze for cars will be one step ahead of its rivals if it knows in advance of a change to colder weather and has plenty of supplies.

Weather forecasts are vital for shipping; they particularly emphasise wind strength and direction as well as information on visibility. Gale warnings are issued separately when necessary. A ship's captain can use weather information to decide on a route that avoids bad weather. This may result in a saving of fuel and increase the safety of passengers and cargo. You may be familiar with these forecasts as they are supplied by the Meteorological Office and broadcast over BBC radio.

Farmers may need to time the beginning of harvesting to avoid a spell of wetter weather or the threat of frosts (Fig. 1.4). The Meteorological Office has a specialist department dealing with the particular needs of farmers and growers.

Aviation is the largest consumer of weather information and meteorology also forms an important part of a pilot's training. Pilots have to be able to recognise potentially dangerous weather conditions that could pose a threat to the safety of their aircraft and passengers. Satellites can now transmit details to aircraft of the position of fast eastward-moving high-altitude air currents called **jet streams**; a pilot will often alter course to find these jet streams. Airline pilots are given a lot of information at group briefings before flights. This includes details of jet stream positions, for these are constantly changing, and how they should influence flight time (Table 1.2). For flights from the USA and Canada, about six to eight routes are offered which make use of the jet streams (see section 6.5) on eastward journeys.

4 Study Table 1.2 and suggest why airline pilots use jet streams when flying eastwards to the UK from the USA.

5a Identify some of the Meteorological Office's major markets for weather information from their publicity material in Figure 1.5. Suggest in each case why information might be needed by these markets.
b Think of other markets which may require the Meteorological Office's services.

Table 1.2 Selection of planned flight times from London and Manchester airports, 9 and 10 August 1994 (*Source:* British Airways)

Airports	Flight time: west-bound	Flight time: east-bound
Heathrow/Los Angeles	11 hours	10 hours, 20 minutes
Manchester/New York	7 hours, 6 minutes	6 hours, 13 minutes
Gatwick/Atlanta	8 hours, 21 minutes	7 hours, 43 minutes

The Meteorological Office in Britain has traditionally been a weather forecasting agency. Today it has a much more commercial role supplying environmental and business advice related to the weather (Fig. 1.5).

There are links between the weather and the way in which noise travels through the air. A computer program developed at the University of Salford, in conjunction with the Ministry of Defence, is being used to predict when noise pollution might be at its worst during certain atmospheric conditions. If, for example, a quarry was planning to carry out some blasting and the program indicated that weather would lead to sound being carried towards a built-up area, then operations could be delayed. Further examples of applications for this technology might include noise control near airports and army firing ranges.

Figure 1.5 Selection of markets for the Meteorological Office (*Source:* Met. Office)

The Daily Telegraph, 9 April 1993

Avalanche warnings as blizzards sweep slopes

The Daily Telegraph, 27 November 1993

Fog brings chaos to the motorways

Storms flatten pecan trees in Georgia

The Independent, 26 November 1992

The Daily Telegraph, 4 February 1993

Spain prays for rain to end drought in the south

16 die as tornadoes hit US

The Independent, 23 November 1992

Figure 1.6 The impact of the weather

At a different scale, atmospheric processes can often become hazardous (Fig. 1.6). Sometimes exceptional events are reported like the storms leading to the Mississippi floods in 1993, or hurricane damage leading to widespread destruction and deaths.

Clearly we need to understand the workings of the atmosphere, not only to provide data for contemporary applications but also to resolve future problems. People interact with the atmosphere: we do not just passively receive the weather but also, through our activities, we modify atmospheric processes. Two of the greatest environmental problems at the end of the twentieth century are global warming and damage to the ozone layer; both are largely the result of human activities.

In this book you will find examples of how geographers are using their skills to synthesise and analyse information from a variety of sources. This can then be used in a predictive way, or in a management context, to reduce the impact of hazards on communities in different parts of the world.

1.3 Data collection and the role of satellites

For most of human history people have turned to religion for explanations of the weather. Although there were a few early attempts to describe and explain atmospheric conditions, for example, Aristotle's *Meteorologica* in the 4th century BC, there was little advance in understanding until meteorological instruments were developed in the seventeenth century. The invention of the barometer in 1643 by Evangelista Torricelli represented the beginnings of a scientific approach to meteorology. By the end of the seventeenth century, pressure, temperature and humidity could all be measured.

Although the first recorded weather map was made in 1686 by Edmond Halley, it was not until the telegraph was invented in the 1840s that people could produce effective weather charts. Improvements in communications meant that information from different weather stations could swiftly be sent to one centre where a picture of the weather over a wide area could be produced (Fig. 1.7). However, many problems remained. For example, data was not available for much of the earth: over 70 per cent of the earth's surface is oceans and seas. There were also gaps for remote areas such as ice sheets, mountains and deserts. We now know that many processes fundamental to understanding the workings of the atmosphere take place in these areas. In the latter half of the twentieth century, this problem was largely solved with the development of weather satellites. These observe and measure atmospheric processes over the whole globe.

_____ **?** _____

6 Suggest how the weather might influence the following events:
• the British Grand Prix, Silverstone,
• the Chelsea Flower Show, London,
• the Grand National, Aintree,
• the Farnborough Air Show, Surrey.

7 Compile a table of different climatic hazards. In one column include brief statements about groups of people or human activities that might be particularly affected. In a second column include any specific recent examples referring to dates and locations. You could use backdated newspapers on a CD ROM to complete this exercise.

Figure 1.7 Synoptic chart based on data gathered by telegraph, 25 October 1859

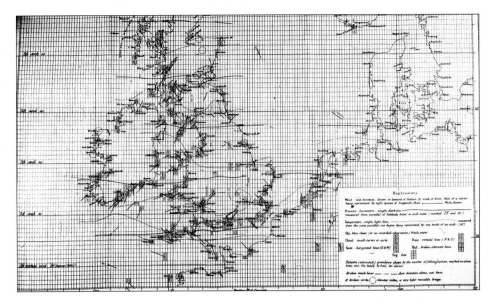

Figure 1.7 Synoptic chart based on data gathered by telegraph, 25 October 1859

Figure 1.8 Satellite image of Hurricane Allen over the Gulf of Mexico, 8 August 1980. The data for this image were gathered by a Geostationary Operational Environmental Satellite.

Satellites in meteorology

For thousands of years people could only look at the weather from the earth's surface. For a few decades we have been able to ascend into the atmosphere, but only since the first weather satellite was launched in 1960 have we been able to look down on weather systems from space (Fig. 1.8). This led to huge advances in weather forecasting, as well as enhancing our understanding of many atmospheric processes and aiding our monitoring of climate change.

Although satellites have in the past been launched by rockets, increasingly the space shuttle has been used to position satellites in their orbit and to carry out servicing. The time taken for one orbit depends on its height above the earth's surface. The altitude of the orbit can be decided before the satellite is launched. European countries, including the UK, currently receive data from two types of satellite.

NOAA

Two NOAA (National Oceanic and Atmospheric Administration) satellites operated by the USA orbit the earth from pole to pole (Fig. 1.9), crossing the equator in about 1 hour 42 minutes at a height of about 670 km. During this time, the earth moves in its rotation about 25°, so a different part of the earth is scanned on each successive

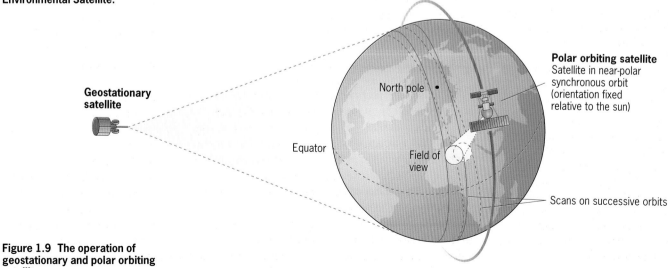

Figure 1.9 The operation of geostationary and polar orbiting satellites

11

Figure 1.10 False colour satellite image of Europe. This image was made in visible light by the NOAA 11 satellite, 4 July 1991

Figure 1.11 False colour satellite image of Europe and North Africa. This image was made in visible light by the Meteosat, August 1989

8 Study the two satellite images in Figures 1.10 and 1.11. Comment on the degree of detail in the cloud cover in each image.

9 From Figure 1.12, what can you conclude about the degree of cloud cover over the centre of the British Isles in contrast with the area to the west of France?

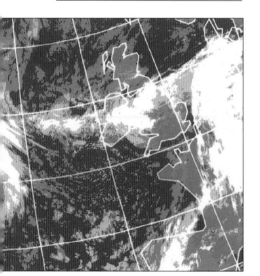

Figure 1.12 NOAA infra-red satellite image of western Europe: thin low-lying cloud is dark grey and thicker, higher cloud is light grey/white; land green, water blue, 12 September 1994

orbit. Because these *polar-orbiting satellites* have a lower orbit, the images give detailed pictures of cloud cover (Fig. 1.10). These images are transmitted to the UK twice a day from three passes, one over the eastern Mediterranean, another over the UK and the third over the eastern Atlantic.

Meteosat

Meteosat is a *geostationary satellite*. This means that it orbits in space but remains in the same position above the equator on the earth's surface. In its orbit at a height of about 36 000 km, Meteosat takes 24 hours to travel round the earth. These satellites continuously scan one large area, as shown in Figure 1.9. Meteosat is operated by European countries and provides images of the North Atlantic and Europe every thirty minutes. These images are not very detailed because of the altitude of this satellite (Fig. 1.11).

Creating satellite images

Satellites look at the earth and measure radiation with instruments called *radiometers*. Moving mirrors enable the satellite to view different parts of the earth in a pattern known as a *scan* (see Fig. 1.9). A series of scans is used to build up a complete picture and the data are then transmitted to an earth station using a radio link.

One type of radiometer works rather like a conventional camera: it provides images based on measured visible light. If sunlight is being measured which has been reflected from the upper surfaces of clouds, it follows that thick cloud will be shown as white areas and thinner cloud in shades of grey (Fig. 1.10). One problem is that these images are only available during the day.

Radiometers can also measure infra-red radiation that relates to the temperature of clouds and the earth's surface. Figure 1.12 shows such an image: the colder areas generally appear white and warmer areas grey. If temperatures broadly fall with height in the lower atmosphere where clouds are found, it follows that high cloud will appear white and lower cloud in darker colours. Unlike visible light images, infra-red images can be produced at night. Care needs to be taken not to confuse warm cloud and the ground surface or high, cold cloud and areas of ice.

Europe suffers flood damage

Thursday night's storm hit Spain earlier than the rest of Europe, reaching its greatest intensity there late last Wednesday night and Thursday morning, after two days advance warning from the Spanish Weather Bureau.

The main problems were caused by heavy rainfall and flooding, and one person is known to have died.

Figure 1.13 Storm warnings given by the Spanish Weather Bureau (*Source: The Times*, 17 October 1987) © Times Newspapers Ltd., 1987

10 Figure 1.13 refers to a storm warning given in October 1987 in Spain. The Meteorological Office in the UK was severely criticised for inadequate warning before the devastating storms that affected southern Britain a few hours later. If people have advance warning of a powerful storm, what steps could they take to reduce the amount of damage to life and property?

11 Essay: Describe and assess the role of satellites in meteorology.

Using satellite images

Satellites have helped to provide information about areas of the globe where previously few observations were made e.g. major oceans and polar areas. Satellites can now give regional and global coverage of weather systems, monitoring their location and movement. Meterologists have been able to use such data to explain the distribution of different climate types and climate change. Other uses of satellites include:

- Measuring the temperatures of the sea surface and cloud tops.
- Measuring snow and ice cover.
- Producing vertical temperature profiles of the atmosphere and measuring the concentrations of gases such as ozone.
- Measuring the movement of depressions for forecasting purposes.

Satellite images can also be used to reduce the impact of weather on people, sometimes preventing a natural event from becoming a disaster. They can show the build up of an intense depression and suggest that gale warnings be issued. This frequently happens in Britain: when there are warnings of strong winds or blizzards, people can take preventative action (e.g. avoid travelling by car) and the scale of damage is thus reduced. It is possible to use satellite images to monitor hurricanes and predict their tracks. Subsequent evacuation of areas in their path can reduce loss of life. Satellite images can also show changes in the position of ocean currents. Meteorologists can then estimate the possible climatic implications of such change before the event. This may enable effective planning for a period of drought (see section 3.4). Satellites can provide data so rapidly that prediction has become more accurate and forecasts give information that is of great value to many people like farmers, sailors and those working on oil production platforms. Later chapters use satellite images extensively to illustrate various atmospheric processes and weather systems and to help you to appreciate their importance more fully.

1.4 Atmospheric processes and weather forecasting

Models of the atmosphere

Weather forecasting and our understanding of the atmosphere have traditionally been based on the assumption that atmospheric processes follow a set of laws. Our improved understanding of the atmosphere is not just the result of a massive increase in the quantity of data or improved coverage of the planet. Many theories and proven atmospheric processes are now converted to mathematical relationships. These in turn provide an aid to forecasting and monitoring atmospheric change (see section 3.5). Computers process the enormous quantities of data and quantify the relationships between the many variables involved. Meteorologists have developed sophisticated climate *models* (general circulation models, or GCMs) to simulate the workings of the atmosphere (Fig. 1.14). A model is a simplified version of reality and can take various forms, for example, a computer program, a diagram or a mathematical formula.

A systems approach to meteorology

Although a systems approach is used widely in this book, it is not the only analytical method, nor always the most appropriate. However, it is a valuable tool to aiding our understanding of many of the complexities of the atmosphere.

To understand the atmosphere, it is necessary to look not only at the atmosphere itself but also how it interacts with space and the earth's surface. Broadly, there are three main flows in the atmosphere: movements of energy, water and gases. To understand the workings of the atmosphere is clearly an immense problem.

Figure 1.14 Studying an atmospheric computer model

Systems theory is one approach. It represents an effort to generalise reality and to break it down into the main components or elements and the linkages between them.

Figure 1.15 shows a simple view of the earth, its atmosphere and space as a system. This is a simple diagram which shows *inputs* and *outputs* but does not give us a view of any internal processes that might explain the links between them. A more familiar example of a system is the water circulating in a house (Fig. 1.16). Water, an input, enters a house and is stored in a boiler or a tank in the roof. Gravity then sends water in pipes around the house to the taps and toilets. After use, waste water, an output, leaves the system and passes into the drains and sewers. This is an example of an *open system*, one that is characterised by inputs and outputs of matter and energy across the boundaries of the system. The production of hot water for central heating or other domestic uses provides examples of *sub-systems* (Fig. 1.17). These take place within the system boundary shown in Figure 1.16.

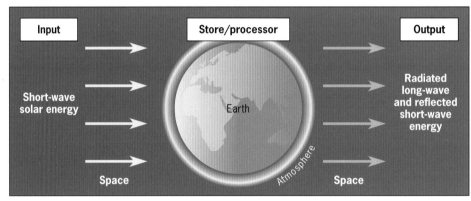

Figure 1.15 Systems view of earth, atmosphere and space

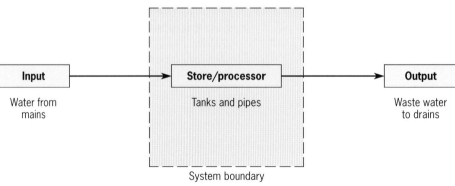

Figure 1.16 The water system in a house

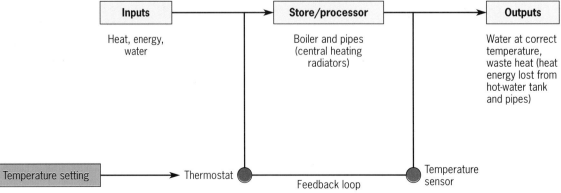

Figure 1.17 The water heating sub-system in a house

Feedback

In Figure 1.17, water is heated either in a boiler or an immersion heater. Thermostats are designed to prevent overheating and control the temperature of the water. A sensor responds to the water temperature and activates the thermostat when the temperature rises or falls below a certain level. This is an example of *negative feedback* which *reduces* the effect of changes so that the system maintains a *dynamic equilibrium* (Fig. 1.18). If the sensor or thermostat failed, heat energy would continue to enter the system until it over-heated and broke down. This is an example of *positive feedback* (Fig. 1.19), which *increases* the effect of changes so that the system is incapable of remaining in equilibrium. Later chapters provide a number of examples of both types of feedback.

Figure 1.18 Negative feedback path in a domestic hot water system (below)

Figure 1.19 Positive feedback path in a domestic hot water system (below right)

Fog

Figure 1.20 represents the application of the systems concept to fog, a familiar weather phenomenon. Positive feedback operates both during the night and early in the morning to encourage the development and dispersal of fog.

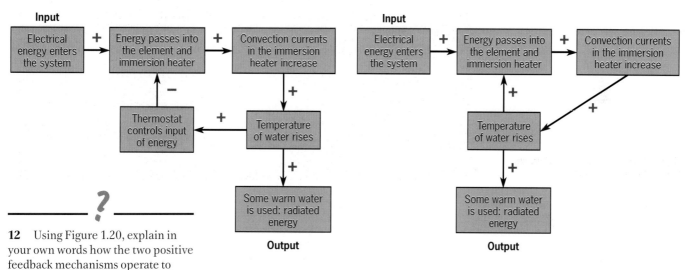

12 Using Figure 1.20, explain in your own words how the two positive feedback mechanisms operate to encourage the development and dispersal of fog.

13 Use Figure 1.20 as a guide and complete a flow diagram showing negative feedback.
a Place the following labels into boxes in the correct sequence to show how fog can reflect **insolation** (solar energy) to compensate for the effect of rising amounts of insolation.
• OUTPUT: reflected energy
• Encourages evaporation of water
• INPUT: insolation as sun rises in the morning
• Bright upper surface of fog reflects insolation
b Add positive and negative signs beside each arrow to indicate the relationships between each component in the system.
c What do you think will happen to the system as the sun rises and the level of insolation increases?

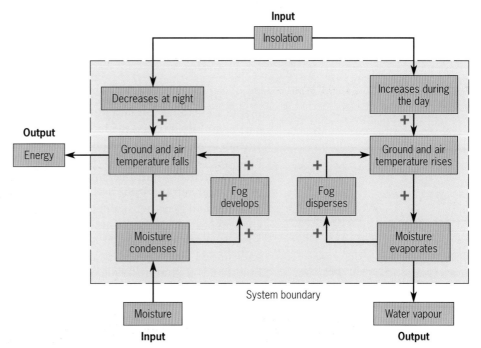

Figure 1.20 Positive feedback: the development and dispersal of fog

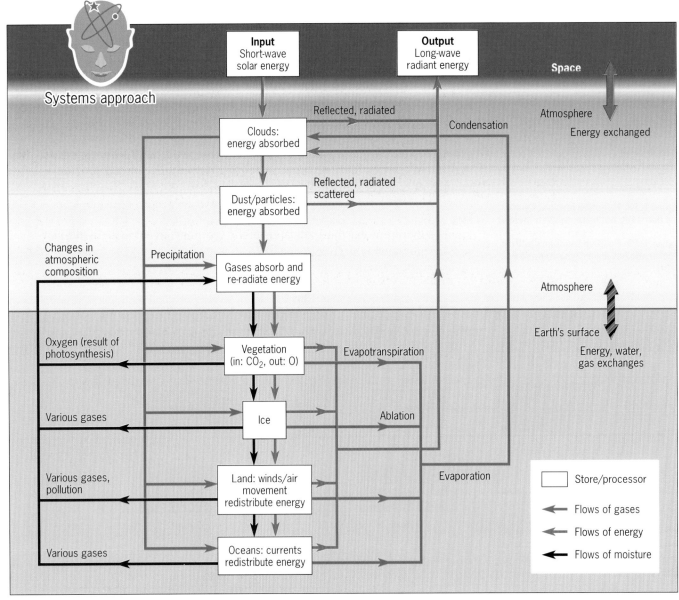

Figure 1.21 The main elements of the atmosphere as a system

Flows of energy and matter

Figure 1.21 is an attempt to summarise in one diagram the main elements of the earth's atmosphere and the interactions between them. This is a more complex version of Figure 1.15 that allows us to see inside the earth/atmosphere system. It is now easier to comprehend how the system works, although the complexities in all their detail may require a computer to handle or model the interrelationships.

Although the atmosphere is an open system as far as energy is concerned, moisture and gases neither enter nor leave the system: there are only flows and changes of state within the atmosphere-earth system (Fig. 1.21). Incoming short-wave solar energy crosses the space-atmosphere interface and reflected short-wave and emitted long-wave energy radiated from the planet passes back to space. These inputs and outputs are balanced over a long period and reflect many feedback mechanisms.

Finally, Figure 1.21 also gives some indication of the role of people in the system and how, both intentionally and unintentionally, they affect atmospheric processes. There are many examples in later chapters of human disturbance which occurs at such a scale that the system is not able to adjust quickly enough.

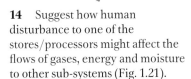

14 Suggest how human disturbance to one of the stores/processors might affect the flows of gases, energy and moisture to other sub-systems (Fig. 1.21).

15 Why is a systems approach useful in meteorology?

Chaotic systems and forecasting

The atmosphere is made up of many sub-systems that interact with each other (Fig. 1.21). At a large scale and over time, these interactions produce climates that are fairly stable. However, at a small scale and over short periods, the interactions produce weather phenomena that are chaotic or unstable, such as winds, clouds and precipitation. In short-term weather events, patterns do repeat themselves, but never in exactly the same way, which makes accurate short-term forecasting very difficult.

Chaos theory explains this by saying that weather is very sensitive to changes in initial conditions: tiny changes to the inputs of gases, energy and moisture in the atmosphere can have a large effect on a weather system. For example, a sudden drop in temperature could start a chain of events that leads to a hurricane. So, although computers can now cope with large amounts of data very quickly, chaos theory suggests that weather forecasting will never be a precise science but an approximation of what can be expected.

Weather forecasting

Weather forecasts provide useful information to industry, agriculture and commerce. In addition many members of the public regularly follow forecasts from the media. There are three basic steps involved in making a forecast:
• observation, • analysis of the data collected and • preparation of the forecast from material provided by computers.

Collecting the data

Figure 1.22 shows some of the many different sources of data from around the world. Satellites and automated ground stations are increasingly adding to, and

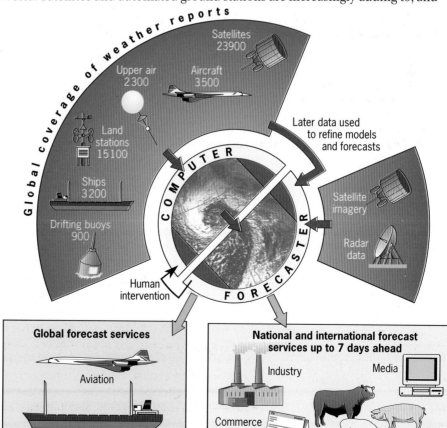

The numbers give an indication of how many observations are used in each 24 hour period as a basis for the global forecasts.

Figure 1.22 The flow of information from observations to forecasts (*Source:* Met. Office)

Figure 1.23 Weather instruments at Ringway, Manchester

Figure 1.24 Principal weather stations in the UK (*Source:* Met. Office)

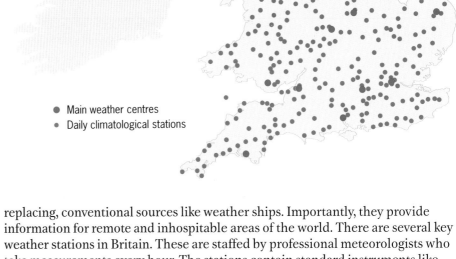

● Main weather centres
· Daily climatological stations

Figure 1.25 Radar composite of rainfall over the British Isles, 1 June 1992. Heaviest rain is shown in red, graduating to yellow, then green for lightest.

?

16 Study Figure 1.25. Describe the pattern of rainfall over the British Isles indicating which areas are experiencing heavy and light rainfall.

replacing, conventional sources like weather ships. Importantly, they provide information for remote and inhospitable areas of the world. There are several key weather stations in Britain. These are staffed by professional meteorologists who take measurements every hour. The stations contain standard instruments like barometers, rain gauges, anemometers and thermometers (Fig. 1.23). However, forecasters need additional spatial information to enable them to compile a more thorough and accurate picture of the weather. This is provided by several hundred other sites around the country (Fig. 1.24). Some of these are staffed by personnel of the Meteorological Office, while others are operated by observers such as the coastguard. At sea there are over 500 ships and rigs forming part of the UK Voluntary Observing Fleet. A larger scheme involves over 7000 ships belonging to 49 nations. There are also a number of automated buoys to the west of Ireland in the Atlantic.

Information about the upper atmosphere like temperature, pressure and humidity is collected by balloon-borne instruments called radiosondes. These are released on a tether, or cable, and collected twice a day at a network of centres around the world. Mini-sondes can be sent up when required to obtain additional information. Aircraft reports of temperature and winds can also add to this data.

Table 1.3 Some costs and benefits of forecasting

Costs
- Investment in different forms of data collection: satellites, weather ships, aircraft, ground recording stations etc.
- Investment in computers: hardware and software, including supercomputers (among the most expensive in the world).
- Investment in communications systems, both within and between countries.
- Inaccurate forecasts which may cost: loss of professional reputation for Met. Office, loss of lives and damage to property, loss of crops etc.

Benefits
- Lives saved and property protected due to accurate forecasting and advanced warning of extreme conditions.
- Increased safety for people working in hazardous environments e.g. oil platforms.
- Crops harvested in time to avoid strong winds, heavy rain, hail or frost etc.
- Fuel saved by ships as they avoid areas with strong winds or storms.
- Fuel and time saved by airliners flying eastwards as they successfully find and fly in jet streams.
- Improved planning for many events e.g. horse and motor racing, garden parties, air shows etc.
- Shops can purchase strategically to coincide with changes in the weather, and so benefit their incomes.

A UK network of weather radar provides a picture of rainfall patterns. We often see radar images of these on television weather forecasts. By using radar, it is possible to detect areas of rainfall and how heavy it is (Fig. 1.25). Radar information from Europe is also available. Satellites have been one of the most significant sources of information since the 1960s. In Chapter 9, which deals with mid-latitude climates, we make extensive use of both weather maps and satellite images.

Analysing the data
The data collected by these various means (and forecasts) are transferred around the world by the Global Telecommunications System (GTS). This comprises a series of computer-to-computer links using satellites and land lines. The GTS centre in Bracknell, the home of the Meteorological Office, passes data between Washington (USA) and continental Europe. The observations taken from the GTS are stored on computer and can then be used in two ways.

Firstly, computers plot the observations for a particular time on a chart. Forecasters then analyse these charts with the additional information from radar and satellites to understand what is happening to the weather. The compilation of a chart involves plotting **isobars**, adding (after identification) depressions or anticyclones, and other information such as wind speed and direction, cloud cover and precipitation.

Secondly, observations are also entered into a computer. The computer reduces the atmosphere, the sea, the sun's warmth and the spinning of the earth to mathematical formulae. This model can accurately reproduce the seasons, wind patterns and, of course, changes in the weather.

Forecasting the weather
The final stage is forecasting, and here the computer models play a key role. In about 15 minutes the 'super computer' at the Meteorological Office in Bracknell can produce a six-day global forecast. This is produced twice a day using observations at midnight and midday as the starting point. Every day the computer model also forecasts the weather for the next few days over Europe and the north-west Atlantic. Forecasts of all the weather elements are given, and we often see those of surface pressure, wind, temperature, cloud and rain in the media. Contrary to popular belief, the computer models do work! Hemispheric, zonal and regional averages are well produced and unusually hot or cold years are predicted. Small-scale averages and changes are not so well simulated; precipitation for example is particularly poor in certain regions. The models are being improved, though, as more is learnt about atmospheric processes. The Hadley Centre for Climate Prediction and Research in Bracknell, Berkshire, is one of six in the world that combines the 'undersea' climate with that of the atmosphere above (see section 2.5).

Human weather forecasters are still needed, though, for several reasons. They can often use local knowledge and experience to fill in details that the computer with its broader spatial coverage cannot provide. Late observations and the latest radar and satellite information mean that forecasters can add fine details to their forecast that may be very important for particular customers. Farmers, for example, need to know when early frosts are likely to occur in the autumn. One of the Meteorological Office's largest customers is the North Sea oil and gas industry. Accurate information on wind speed and strength has particular implications for the day to day operation of oil and natural gas platforms (Fig. 1.26).

Forecasting is very sophisticated today but it can still be limited by the weather conditions at the time of the forecast. When the weather is changeable it is possible to predict 24, perhaps up to 72 hours ahead. However, during anticyclonic conditions accurate forecasts of up to a week are possible. One top weatherman said '. . . despite computers and satellites, long-range forecasting is still an uncertain business . . . There's no hotline to heaven for guidance.'

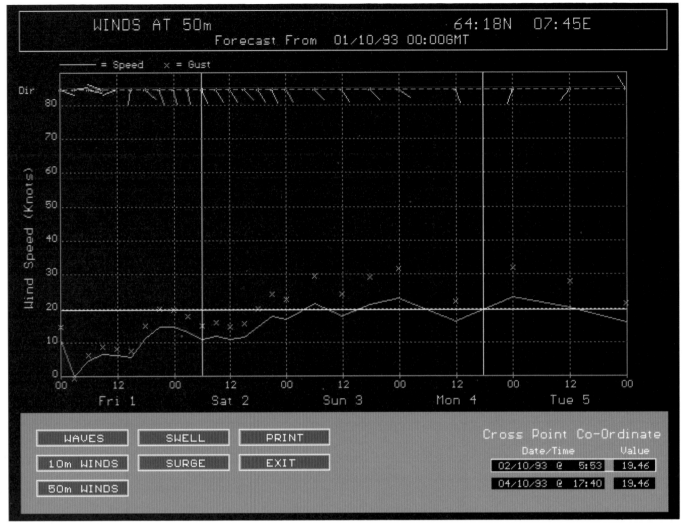

Figure 1.26 Five-day ahead Met. Office computer simulation forecast for windspeed at oil rig level, North Sea. Such predictions are important for deciding whether rig cranes should operate, or helicopters can land. The green line shows the wind speed, crosses represent the strength of gusts while the yellow lines at the top of the diagram show the direction of wind.

Summary

- The atmosphere is constantly changing.
- The terms weather, climate and meteorology have precise meanings that need to be understood.
- There are many two-way links between people and atmospheric processes. The atmosphere influences human behaviour and our activities in turn affect the atmosphere.
- Understanding weather and climate is important as it affects people's lives in many different ways. A number of occupations rely on up-to-date weather forecasts.
- We need to understand atmospheric processes to be able to predict and adapt to climatic changes, such as global warming.
- Satellites and climate models have been of great value in enhancing our understanding of atmospheric processes, improving the accuracy of weather forecasting and aiding our monitoring of climate change. Computers have been vital to store and process the volume of collected data.
- Systems theory is a useful tool to aid our understanding of the complexities of atmospheric processes and how the atmosphere interacts with space and the earth's surface.
- Geographical skills can be used to synthesise and analyse information from a variety of sources. This can then be used to increase our understanding of the atmosphere and applied to reduce the impact of hazards on communities in different parts of the world.

2 The atmosphere: energy in the system

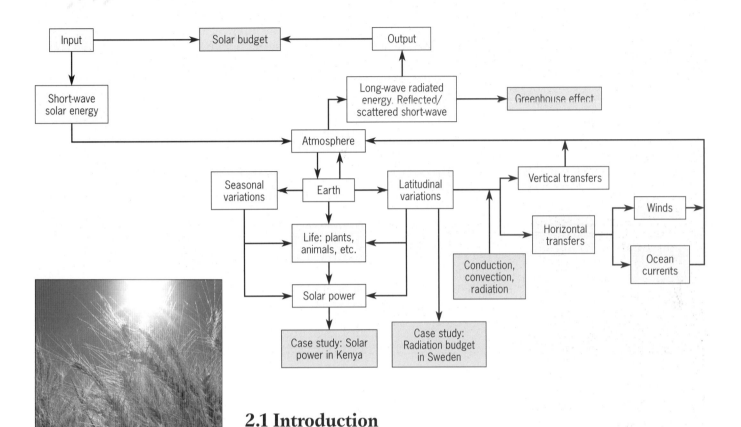

2.1 Introduction

The sun is the ultimate source of virtually all the world's energy. The only other possible contribution is made by geothermal energy, but this is negligible in comparison. What happens to the sun's energy as it passes through the atmosphere has many implications for life on earth. Plants, or *producers*, that are the first stage in food chains represent *stores* of solar energy (Fig. 2.1); this energy is fixed by *photosynthesis*. Herbivores consume plants. In turn, other *consumers*, including people, eat both plants and consuming organisms. Simply, human existence is dependent on the availability of the sun's energy.

The atmosphere has often been compared to an engine, with its operation producing **climate** and **weather**. The fuel for this engine is energy from the sun.

Figure 2.1 The sun's energy (above) begins the food chain (below)

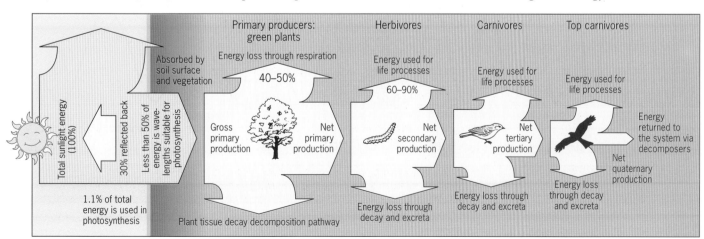

How solar radiation drives the engine, though, is very complex and involves physical, chemical and biological processes.

Dependent as we are on the sun, atmospheric changes brought about by human activity are beginning to upset the way in which the earth receives this energy. Some researchers suggest that **global warming** and the damage caused to the **ozone layer** are some of the greatest threats to the future of life. In this chapter we will examine the nature of the atmosphere, the movements of solar energy within it and spatial and temporal variations in the amount of **insolation** received by the earth. This will be done at the global and regional scale. We will also consider the value of solar energy as a natural resource. Geographical research into the spatial and temporal availability of solar energy is essential if we are to develop fully its potential as an alternative energy source.

2.2 Solar energy as a renewable energy source

The sun is a gaseous sphere with a surface temperature of nearly 6000°C. It emits radiant energy or insolation which travels in waves. The sun releases energy during a nuclear process in which hydrogen is converted to helium. Remarkably, the earth only receives one two thousand millionths of the energy from the sun; the rest radiates into space. This huge amount of radiant energy has no mass. It travels freely through space and, although the visible portion of this energy can be seen, it can only be sensed when it reaches the gases and tangible parts of the earth.

We can describe electromagnetic radiation from the sun by its frequency and wavelength (Fig. 2.2). Solar energy is largely of very short wavelengths. Although some of this energy is reflected from the surface of clouds, most passes through the atmosphere without being absorbed and then warms the land and oceans. About seven per cent of insolation is in wavelengths lower than 0.4 μm (micrometres), including *ultraviolet* energy. This is important for maintaining the ozone layer as it is the breakdown of O_2 into two oxygen atoms by UV radiation that leads to ozone (O_3) formation. At longer wavelengths, the sun emits *infra-red* and radio-wave energy.

From a human point of view the sun is also an inexhaustible source of energy (Fig.2.3). Unlike fossil fuels and nuclear energy, it is clean, renewable and available everywhere. However, despite these obvious advantages, solar energy is not uniformly distributed either spatially (Fig. 2.4) or in time.

Figure 2.3 Solar power station, Albuquerque, USA. Rows of large mirrors (heliostats) mounted on movable frames track the sun and concentrate its light to a central receiver. Here pressurised fluid is heated until it vaporises and drives a turbine generating electricity.

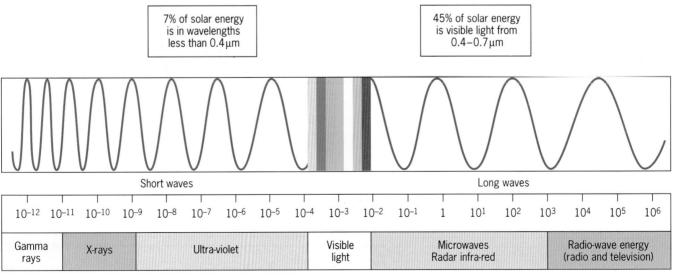

7% of solar energy is in wavelengths less than 0.4 μm

45% of solar energy is visible light from 0.4–0.7 μm

Short waves

Long waves

| 10^{-12} | 10^{-11} | 10^{-10} | 10^{-9} | 10^{-8} | 10^{-7} | 10^{-6} | 10^{-5} | 10^{-4} | 10^{-3} | 10^{-2} | 10^{-1} | 1 | 10^1 | 10^2 | 10^3 | 10^4 | 10^5 | 10^6 |

| Gamma rays | X-rays | Ultra-violet | Visible light | Microwaves Radar infra-red | Radio-wave energy (radio and television) |

10^{-9} = millionths of a millimetre   1 = one millimetre 10^3 = tens of millimetres

Figure 2.2 The electromagnetic spectrum

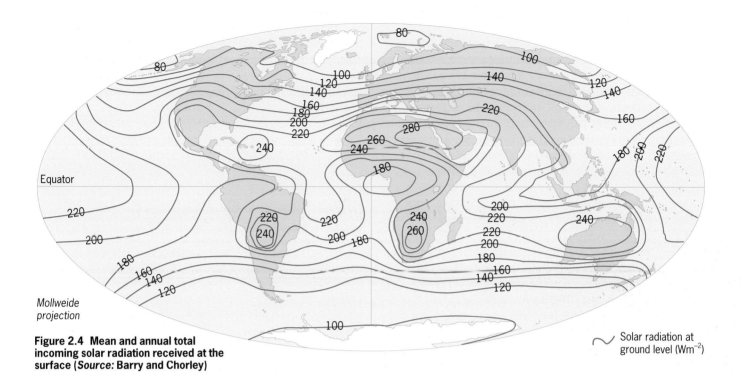

Figure 2.4 Mean and annual total incoming solar radiation received at the surface (*Source:* Barry and Chorley)

Solar radiation at ground level (Wm^{-2})

1a Study Figure 2.4. Describe the pattern of solar energy received at the earth's surface. Use an atlas to name specific areas that either have particularly low or high values.

b Account for the pattern you have described.

The amount of solar energy received at the earth's surface varies according to latitude and the characteristics of the atmosphere. Latitude affects not only the altitude of the sun in the sky but also the spatial concentration of the sun's energy (Fig. 2.5) and the duration of insolation (see section 2.4). Characteristics of the atmosphere like water vapour, turbulence, cloud cover, carbon dioxide (CO_2) and particulate concentrations vary spatially and temporally and influence solar energy as it passes down through the atmosphere. It is important, therefore, that we determine the amount of solar energy received at any point so that we can assess the potential of solar energy as a power source.

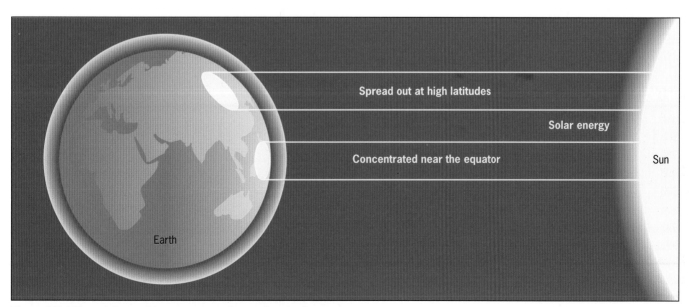

Figure 2.5 Effect of the earth's curvature on the receipt of solar energy

The potential for harnessing solar energy in Kenya

Nearly two-thirds of the world's rural inhabitants have no access to electricity. Rural communities are often widely scattered and remote. They therefore need independent energy supplies that are long-lasting, reliable, simple to maintain and where the 'fuel' source can be taken for granted. Solar energy can often meet these needs. Africa is a part of the world with a great deal of potential for solar energy. BP Solar is one company that has worked extensively in a number of east and central African countries.

Due to its location in the tropics (Fig. 2.6), Kenya has plenty of solar energy. Geographical studies show us, though, that this varies spatially and between the seasons (Fig. 2.7). As in many African countries, traditional uses of the sun's energy in Kenya include drying clothes and agricultural crops like maize (drying is a way of preserving food). Since the early 1980s, however, BP Solar has developed a number of projects that harness the sun's energy in several new and different contexts throughout the country (Table 2.1 and Fig. 2.8).

In order to assess properly the potential for solar energy and so develop appropriate solar devices for an area, meteorologists need to take thorough measurements of radiation at several recording stations. Researchers record total amounts of insolation received annually, as well as monthly variations. For planning purposes, they then average these monthly variations and calculate the standard deviation from the mean calculated.

In fact, Ayoade (1990), in a study of the potential for solar energy in Nigeria, recognised that countries face two tasks when embarking on a solar energy programme. One is for geographers and meteorologists to measure and assess the country's potential solar energy resources. They can do this at established weather recording stations. The second task is for engineers to develop appropriate technology to collect and harness energy from the sun and to find ways of applying these methods.

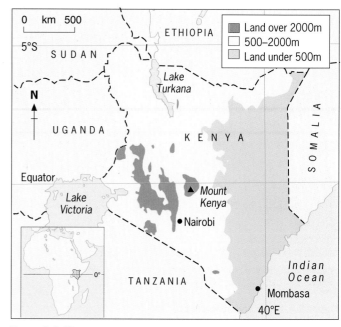

Figure 2.6 Kenya

Table 2.1 Some of BP Solar's projects, Kenya (*Source:* BP Solar)

1985	Health centre lighting
1986	Lighting for Chugu, Njuri and Karaba schools
1986	Complete 240V a.c. power supply for a house
1988	Lanterns for Ker Downey safaris
1988	Medical refrigeration for Chogoria Hospital
1988	Water borehole pump for a game ranch
1991	Radio power system for Kenya Wildlife Services
1992	Medical refrigeration for the Ministry of Livestock
1992	Borehole systems for the Ministry of Livestock (20 to 50 000 litres per day)
1992	Cathodic protection system for an oil pipeline

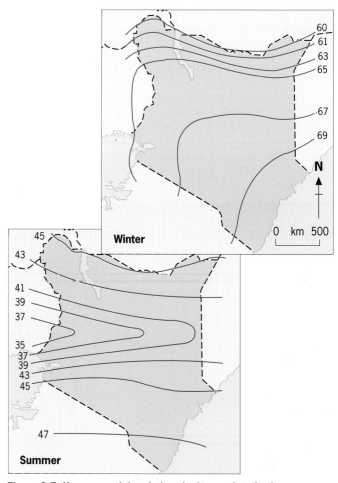

Figure 2.7 Kenya: spatial variations in the receipt of solar energy (*Source:* Sandia National Laboratories, USA)

Figure 2.8 Solar array for school lighting and power, Nairobi, Kenya

2a Study Figures 2.6, 2.7 and Table 2.2. Describe the spatial variations in mean insolation between the seasons in Kenya.
b Give reasons for the variations you have described in answer to 2a.

3a Complete a scattergraph based on the average monthly temperatures and sunshine hours in Table 2.2. Insert a best fit line and give the graph a suitable title.

b Use Spearman's rank correlation to test the relationship for the same data. Test the strength of the correlation using a significance graph.
c Give reasons for the results you have obtained.

4 Suggest some further applications of solar power that could be made in Kenya other than those in Table 2.1.

Table 2.2 Nairobi, Kenya: average temperatures and daily sunshine hours per month (*Source:* BP Solar)

	Jan	Feb	Mar	Apr	May	June	July	Aug	Sept	Oct	Nov	Dec
Average temperatures in °C	18.8	19.6	19.8	19.2	18.1	16.8	15.8	16.2	17.5	18.6	18.3	18.0
Sunshine hours in megajoules/m²/day	20.9	21.5	20.5	17.0	14.6	13.6	12.6	14.2	18.4	19.7	17.0	18.8

2.3 Energy inputs: the role of the atmosphere

The atmosphere can be divided into a number of layers (Fig. 2.9). These have distinct characteristics that partly result from solar energy passing through them. Although the earth's atmosphere is about 1000 km thick, half lies in the lowest 5.5 km, about 99 per cent in the lowest 40 km and at an altitude of 100 km there is

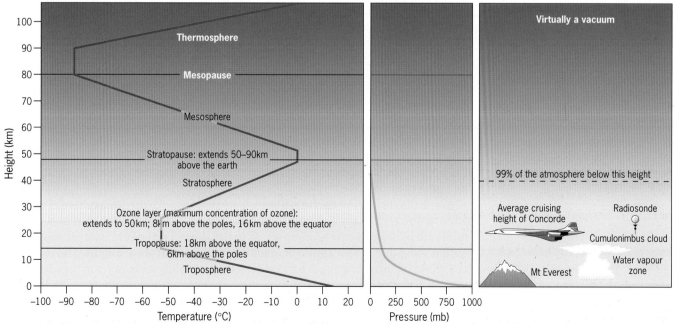

Figure 2.9 Vertical stratification of the earth's atmosphere

Figure 2.10 Dougal Haston using oxygen on the south-west face of Mt Everest

virtually a vacuum. If this is difficult to believe, remember that to reach the top of Mount Everest at 8848 m, many climbers have to wear breathing apparatus due to the lack of oxygen (Fig. 2.10). It follows that there is a very steep **pressure gradient** in the lower atmosphere.

The general decrease in temperature as height increases is partly due to the reduction in atmospheric pressure. As the gases of the atmosphere thin, there are less molecules per unit volume of the atmosphere to absorb solar energy. However, as we shall see, the relationship between temperature and altitude is more complex than this alone suggests.

The atmosphere has weight or **pressure** as a result of the pull of gravity; this is expressed in millibars (mb). The total weight of the earth's atmosphere is 5.8×10^{15} tonnes. The average pressure over the earth's surface at sea level is 1013.25 mb. This is equivalent to the weight of a column of air of 1033.3 g/cm^2. Meteorologists measure atmospheric pressure using either a *barometer* or a *barograph*. A barograph is an instrument that records changes over a longer period of time on special graph paper. This can then be related to changes in the weather (see section 9.1).

Troposphere

Anyone who has ascended in a hot air balloon or climbed in a mountain area will know that the air gets colder with increasing altitude. Temperature changes divide the atmosphere into layers (Fig. 2.9).

The lowest layer, the **troposphere**, contains most of the atmosphere's mass, water vapour and dust. Largely as a consequence of this, most weather processes take place in the troposphere. The **boundary** or **friction layer** is approximately the lowest 1000 m of the troposphere, but this does not have a clearly defined upper limit. Where mountains disturb air flows, this limit may, in fact, extend up to two kilometres. It is in the troposphere that daily localised changes in air movements occur, such as land and sea breezes. Wind speeds generally increase with height because of reduced friction. In contrast, there is a fall in temperature with height as the atmosphere thins. This continues to the **tropopause** – the layer separating the troposphere from the **stratosphere**. The lower levels of the troposphere largely obtain their heat as a result of contact with the earth's surface (see section 2.4).

Stratosphere

The stratosphere, unlike the troposphere, lacks dust and water vapour and is relatively thin. It is also comparatively stable, as the warmer layers lie towards the top. We are particularly interested in the lower stratosphere, though, for a number of reasons, one being that many modern jet aircraft fly at these altitudes. The increase in temperature with height is due to the absorption of solar radiation, particularly ultraviolet wavelengths, by ozone molecules (see section 5.2).

Mesosphere and above

The **stratopause** marks the boundary with the **mesosphere**. Temperatures again fall with height because of the decreasing density of the atmosphere and therefore its inability to absorb energy.

Above 90 km lies the **thermosphere**, synonymous with the **ionosphere**. The atmosphere is virtually a vacuum at these altitudes with pressure about one millionth of the pressure at sea level. As in the ozone layer, there is an increase in temperature due to absorption of radiation at ultraviolet wavelengths. The ions are energised by incoming short-wave radiation that makes them move at very fast speeds. This causes heat sensors to register an increase in temperature with height.

In summary then, the majority of the atmosphere lies in the troposphere, but unlike the other layers it is rarely 'calm'. There are constant changes in pressure, temperature and winds, while weather systems grow, mature and decay.

?

5a Calculate the approximate change in temperature with height for every 1000 m in the troposphere and stratosphere.
b Why do temperatures generally decrease with height?

6 Explain the increase in temperature with height between 13 and 50 km above the earth's surface.

7 Use Figure 2.9 to explain why many jet aircraft like Concorde fly at altitudes in the lower stratosphere.

What is the greenhouse effect?

There is considerable misunderstanding over the terms global warming and the **greenhouse effect**. The warming of the world's climate is commonly viewed as a human problem. The greenhouse effect occurs naturally, though (Fig. 2.11), and is produced by certain gases in the atmosphere. These gases absorb energy and help to warm the earth and maintain temperatures that enable life to exist. As we have seen, short-wave solar energy passes into the earth's atmosphere. Some wavelengths are absorbed, reflected or scattered, while the rest reach the surface and warm the land and oceans. Because the earth's surface is cooler than the sun, the earth emits long-wave, largely infra-red, energy back to the atmosphere where much is absorbed by gases.

There are a number of greenhouse gases but the main ones, in terms of their present contribution to the greenhouse effect, are water vapour and CO_2. Water vapour absorbs infra-red energy in the wave band 4–7μm and CO_2 at 13–19μm. Between 7 and 13μm there is a window through which about 70 per cent of the long-wave energy from the earth's surface can escape and pass back to space. The absorbed energy warms the troposphere and some is reflected back from clouds. This infra-red energy is in turn radiated, some back to the earth's surface, keeping it warmer than would otherwise be the case. The remainder passes upwards through the troposphere and the rest of the atmosphere to escape to space. The release of radiated long-wave energy is broadly in balance with incoming solar energy and is often referred to as the **solar budget** (see section 2.4).

The threat of warming

In the late 1960s when spacecraft first visited Venus, scientists made some disturbing discoveries. The planet's average surface temperature was about 480°C. Initially, scientists could not understand the reasons for these temperatures as Venus is 108 million km from the sun compared with the earth's 150 million km. Venus should therefore *not* be much warmer than the earth. However, some scientists believe that billions of years ago volcanic eruptions emitted vast amounts of CO_2 into the Venusian atmosphere, thus raising the temperature by trapping

Figure 2.11 The greenhouse effect

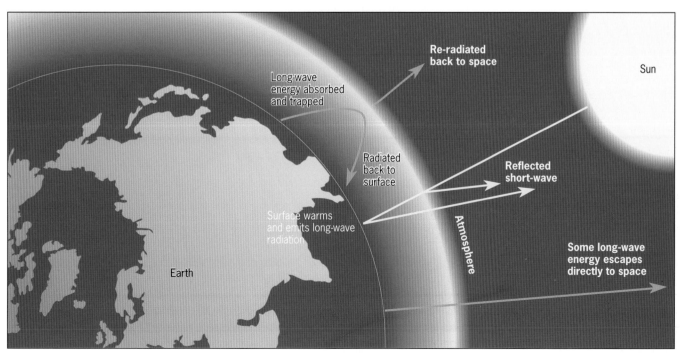

Greenhouse effect

more heat from the sun. This has implications for the changes in CO_2 content on our own planet, whether from natural or human-made sources.

Without the greenhouse effect and an atmosphere, the earth would be like the moon. Here, temperatures rise to over 100°C when lit by the sun and drop to –150°C at night. The average temperature near the surface of the earth would be about –18°C instead of a comfortable 15°C. While the system is in balance, though, conditions remain suitable for life.

2.4 Inputs and outputs: the solar budget

A budget usually refers to an estimate of income (*input*) and expenditure (*output*). People use budgets to balance these two components of income and expenditure. Solar energy budgets involve inputs and outputs of insolation and can relate to different temporal and spatial scales. Energy budgets have important effects on many human activities, particularly agriculture as crops are so sensitive to changes in temperature.

The seasons

The *seasons* are the most obvious consequence of temporal variations in the receipt of solar energy. These result from the changing position of the earth about its axis relative to the sun as it follows its annual orbit (Fig. 2.12). When the sun is directly overhead at the equator (the spring and autumn **equinoxes**) both hemispheres have an equal share of the sun's energy. As the summer and winter **solstices** approach, the earth's axis is no longer at a tangent to the sun but leans away from, or towards, the sun. The overhead sun is now above the tropics and whichever hemisphere receives the most insolation experiences summer.

These seasonal variations in the radiation budget are not just the result of changes in the actual amount of incoming or outgoing energy. They also result from the length of time over which energy is lost or received. During the winter, latitudes north or south of 40° have an energy deficit . This has an important effect on the flora and fauna of these areas as well as on human activities.

Figure 2.12 also shows how there are important differences in the amounts of energy received during the day and the night. These are referred to as **diurnal** variations. You might expect desert areas to have a very positive radiation budget. The deserts in the tropics receive huge amounts of insolation each day and yet they have a similar net radiation budget to locations in temperate latitudes.

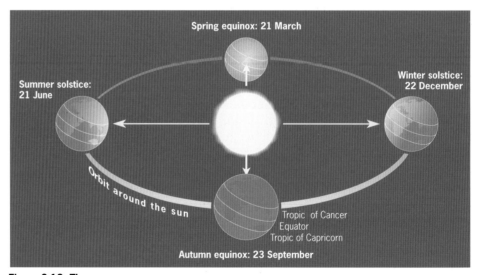

Figure 2.12 The seasons

?

8a Study Figure 2.13. State the latitude at which the maximum annual temperature range for mid-continental areas occurs.
b Give two reasons why it should occur at this latitude.

9a State the latitude at which the maximum diurnal temperature range for mid-continental areas occurs.
b Give two reasons why it should occur here.

10 Give two reasons why the annual maritime range increases towards the poles.

11 Why do equatorial areas not have seasonal variations in their radiation budget?

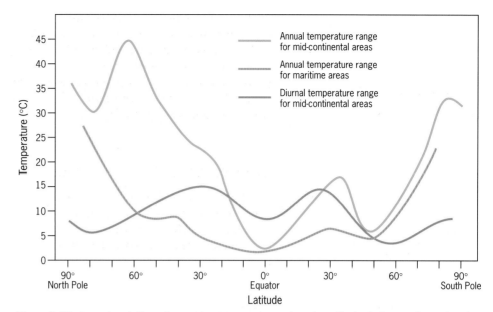

Figure 2.13 Annual and diurnal temperature ranges as a function of latitude for continental and maritime locations

The radiation budget in Sweden

The climate dominates many aspects of life in Sweden (Fig. 2.14). During the summer months in northern Sweden, there is an almost continuous growing season with light and insolation for virtually 24 hours a day. Although plants flourish, producing flowers and seeds and multiplying in number, the climate leads to pronounced latitudinal variations (Fig. 2.15). This is also a time when birds, animals and insects are very active feeding and reproducing. The winters, in contrast, are long and harsh; plants die back and many creatures hibernate or migrate to warmer latitudes.

There is a considerable contrast between the climate in the north and the south (Figs 2.16–2.17). This climatic pattern and energy budget has a strong influence on human activities in the country, particularly where people live. Lapland, the north of Sweden, is a hostile environment for people (Fig. 2.18). Towards the north, winters become progressively long and as the length of snow cover on the ground increases, the growing season decreases. The area offers little potential apart from its forests and minerals. The Lapp people, a minority group of about 15 000, are largely nomadic herders. They move their reindeer from high to lower pastures or the sheltered coniferous forests, depending on the season.

Conversely, much of the south has a density of over 25 people per square kilometre, rising considerably near the three main cities. Although Stockholm experiences a milder climate than the north (Fig. 2.19), its population also has to cope with many difficulties created by the harsh winter weather.

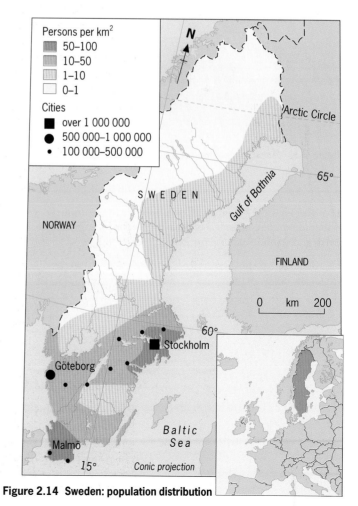

Figure 2.14 Sweden: population distribution

2

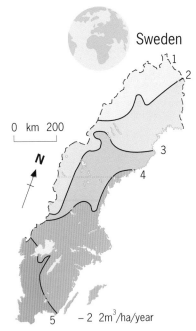

Sweden

0 km 200

N

−2 2m³/ha/year

Figure 2.15 Spatial variations in coniferous forest productivity

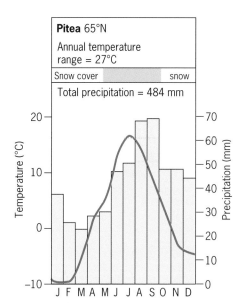

Pitea 65°N
Annual temperature range = 27°C

Snow cover — snow

Total precipitation = 484 mm

Figure 2.16 Pitea, climate

Stockholm 59°N
Annual temperature range = 21°C

Snow cover — snow

Total precipitation = 573 mm

Figure 2.17 Stockholm, climate

Figure 2.18 Snowmobilers stop at a hut to warm up, Lapland

Figure 2.19 Buying Christmas decorations, Stockholm

12 Use Figure 2.14 and an atlas to describe the latitudinal extent of Sweden. Include a comparison of the location of the north and south relative to the British Isles.

13 Explain why northern Sweden experiences an energy deficit during the winter.

14 Study Figure 2.15. Establish a link between the radiation budget for Sweden and the productivity of the coniferous forests.

15 Using Figures 2.14–2.19, list some ways in which differences in the radiation budget between the north and south of Sweden affect human activities.

The solar budget

In this section we consider the gross budget of solar energy for the earth. This is based on total radiation values and ignores (initially) spatial and temporal variations. As the temperature of the earth's atmosphere shows only small changes over long periods of time, it follows that the insolation received must be balanced by outgoing heat. This is called the solar budget. The earth radiates energy of longer wave-lengths than the sun's incoming short-wave energy (Fig. 2.20).

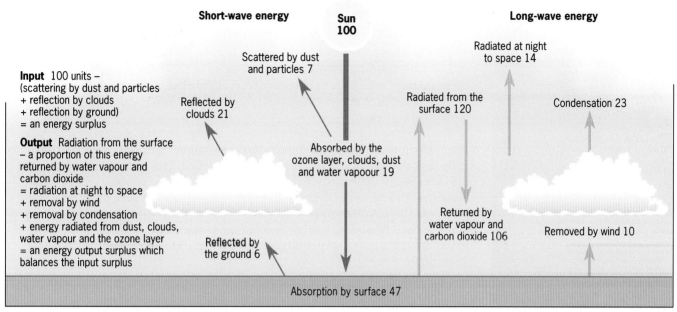

Figure 2.20 The solar budget under average conditions

Input 100 units –
(scattering by dust and particles
+ reflection by clouds
+ reflection by ground)
= an energy surplus

Output Radiation from the surface
– a proportion of this energy
returned by water vapour and
carbon dioxide
= radiation at night to space
+ removal by wind
+ removal by condensation
+ energy radiated from dust, clouds,
water vapour and the ozone layer
= an energy output surplus which
balances the input surplus

Short-wave energy

Sun 100

Long-wave energy

Scattered by dust and particles 7

Reflected by clouds 21

Absorbed by the ozone layer, clouds, dust and water vapoour 19

Reflected by the ground 6

Absorption by surface 47

Radiated at night to space 14

Radiated from the surface 120

Condensation 23

Returned by water vapour and carbon dioxide 106

Removed by wind 10

Insolation

The amount of insolation received at the outer edge of the atmosphere over a period of time, and for a given area, is called the **solar constant**; this is about 1350W/m². The outer edge of the atmosphere reflects a very small fraction of the initial 100 units. Figure 2.20 shows what happens to the solar radiation as it passes through the atmosphere and after it is converted into other wavelengths.

On average, about half of the initial 100 units of energy reaches the earth's surface. Insolation warms the earth's surface to only a few tens of degrees centigrade by means of short-wave energy, largely in the infra-red wave band 4–100 µm (see Fig. 2.2).

Albedo

The earth's surface is very varied in texture and colour. These different surfaces vary in their ability to reflect insolation (Table 2.3). The ratio between the total solar radiation reaching a surface and the amount reflected is called the **albedo**. Albedo is expressed as a decimal or a percentage. The earth's average albedo is about 0.34, i.e. 34 per cent of energy is reflected back to space.

?

16 Study Table 2.3. Why should the albedo for a pine forest (Fig. 2.21) be lower than for a deciduous forest (Fig. 2.22)?

17 Study Figure 2.20. Try to quantify and account for the difference in receipt of solar energy at the earth's surface on a cloudy day and a clear day. Use an annotated diagram to illustrate your answer.

18 Replace the worded equations in Figure 2.20 with numbers from the diagram. Now complete the calculations. If you do this correctly you will find that the input equation produces an energy surplus that is balanced by an output surplus.

19 Explain how global warming resulting from an increase in greenhouse gases would disturb the system and lead to an imbalance in the solar budget.

Table 2.3 Albedo values for selected surfaces

Surface	Albedo (percentage of short-wave radiation reflected)
Dense, dry, clean snow	86–95
Sand – clean	37
Ice sheet	26
Grass	25
Deciduous forest	17
Pine forest	14
Ploughed field – moist	14
Swamp	10–14

Figure 2.21 Oblique aerial view of a coniferous forest, Alberta, Canada

Solar budget

Figure 2.22 Oblique aerial view of a deciduous forest, North Carolina, USA

Figure 2.20 also shows the earth and its atmosphere combined, with energy inputs and outputs in balance. This balance does not, however, apply to the two separate *sub-systems*. Several important points arise from Figure 2.23:

• The earth's surface, taken on its own, has a surplus of energy (radiation inputs exceed radiation outputs) everywhere except near the poles. At these latitudes there are huge losses due to reflection from clouds and areas of ice.
• The atmosphere everywhere loses more energy than it gains, with little latitudinal variation.
• The combined earth–atmosphere system shows two latitudinal zones. Between about 40° north and south of the equator there is a net radiation surplus while polewards of this zone there is a net radiation deficit. This pattern is largely due to the curvature of the earth and the atmosphere relative to the angle of incoming radiation. An equivalent amount of energy is more widely distributed towards the poles and also passes through more atmosphere as the insolation enters at a tangent to the surface. Insolation is more concentrated towards the equator. At this point it enters the atmosphere perpendicular to the surface and also passes through less atmosphere (see Fig. 2.5).

In the absence of some balancing mechanisms two points follow from this: firstly, the earth's surface would be getting progressively warmer and the atmosphere cooler. Secondly, a zone 40° north and south of the equator would be getting warmer, while further north and south would be getting colder.

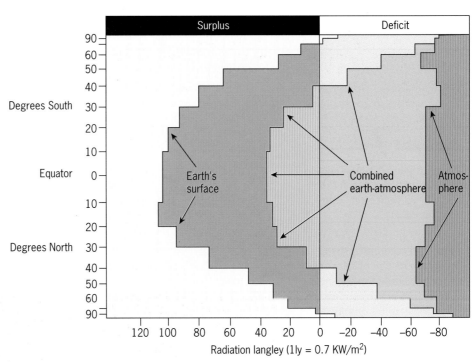

Figure 2.23 Annual net radiation balance of the earth–atmosphere system (*Source:* Hilton, 1986)

2.5 Energy transfers: winds and ocean currents

The system in Figure 2.20 shows only an average picture: there are considerable variations in the energy budget over time (over hours, days and seasons) and over the earth's surface. These variations occur because the inputs of heat energy are not even across the globe and because energy is transferred within the earth–atmosphere system.

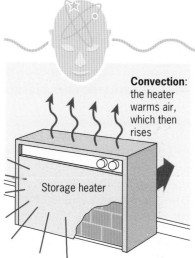

Convection: the heater warms air, which then rises

Storage heater

Radiation: the heater sends out infrared photons

Conduction: hot bricks inside warm the casing, which warms the wall

Figure 2.24
A familiar example of the processes of radiation, convection and conduction

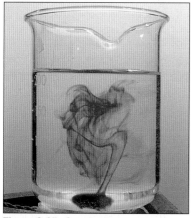

Figure 2.25 Convection currents in water (shown with potassium permanganate)

Conduction, convection, radiation

The processes by which heat energy is transferred are conduction, **convection** and radiation (Fig. 2.24). Conduction takes place through solids such as rock, whereas convection involves movement within liquids and gases (Fig. 2.25). Transfer by radiation does not need the presence of any materials, and that is the process whereby solar energy can reach the earth through space.

Vertical energy transfers

Vertical transfers carry surplus energy from the earth's surface to the atmosphere. This is largely carried out by the conduction of heat to air in contact with land. As warm air is less dense than cold air, it rises by convection. Colder air then moves in to replace the warm rising air, winds also bring air into contact with the ground and through turbulence mix warm and cold air. Most insolation reaching the surface heats the oceans and seas. Considerable amounts of energy (**latent heat**) are required to convert water to vapour through evaporation. When water vapour rises as a result of convection and turbulence, this latent heat is also carried into the atmosphere. When condensation occurs, latent heat is released as sensible heat (which we can feel) to warm the atmosphere.

Latitudinal energy transfers

Latitudinal or *meridional heat transfers* of energy mostly (about 80 per cent) occur through the world's winds, involving conduction and convection (see Fig. 2.25). **Depressions** (areas of low atmospheric pressure) and **anticyclones** (areas of high pressure) transfer a considerable amount of energy in the middle latitudes.

Ocean currents

Ocean currents carry the remaining 20 per cent of energy (Fig. 2.26). These huge flows can extend over thousands of kilometres and affect the climate of nearby land areas (see sections 6.6 and 12.4). The major ocean currents are generated by prevailing winds blowing across the surface. These are influenced by the rotation of the earth and the distribution of the land masses. The consequent pattern of world ocean currents (Fig. 2.27) is essentially the result of the tilt of the earth on its axis, the earth's rotation and the uneven distribution of land and sea areas.

Figure 2.27 reveals that the currents largely follow a huge circular route called a **gyre** in each ocean basin. This moves clockwise in the northern hemisphere and anticlockwise south of the equator. The exception is the circumpolar current that flows around the continent of Antarctica (Fig. 2.26).

Figure 2.26 Computer model of ocean currents. The colours represent vertically integrated velocities from red (fastest) to blue (slowest). The model has predicted many major currents including the Agulhas (bottom left), Kiroshio (upper centre), Falkland (bottom right) and the Gulf Stream (top right). The wind-driven Antarctic Circumpolar Current runs across the bottom of the frame.

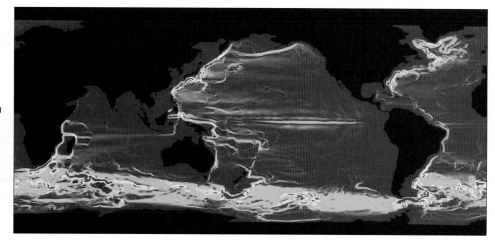

Conduction, convection, radiation

20 Describe and explain the pattern of ocean currents in Figure 2.27.

21 Compare Figure 2.28 with a map of world vegetation regions in an atlas. Comment on the relationships between the patterns.

22a Study Figure 2.28. Describe the pattern of energy balance shown on the map.
b Account for the areas with values above and below 0. Consider in particular why the isolines should bend towards the equator over the land areas.

23 Essay:
a Explain the concept of the solar budget.
b Outline the mechanisms by which energy is redistributed between areas that have a net energy surplus and those with a net energy deficit.

The effect of the currents' circulation within each ocean basin is to pile water into a dome. In the Sargasso Sea, for example, the level of the water in the centre is about one metre higher than at the nearest coast. As the earth rotates from east to west, it also has the effect of piling water up on the western side of ocean basins. Sea level on the east coast of Japan is therefore slightly higher than on the west coast of the USA. As the gyres are forced to the west side of the basins, this in turn encourages relatively fast moving currents, such as the Gulf Stream (Fig. 2.27). Researchers estimate that the Gulf Stream delivers 10^{15} watts of heat – the equivalent of a million major power stations. Currents flowing on the eastern side of the oceans, however, tend to be weaker and more diffuse.

In addition to the surface currents of the world, there is also an *oceanic conveyor belt*, or deep ocean circulation, that corresponds to the atmosphere's climate. Figure 2.27 suggests how important Antarctica is in this pattern of movement. Here, vast amounts of water freeze into ice. This loss of fresh water results in sea water becoming denser and more saline. This denser water consequently sinks and makes its way towards the equator. Vast quantities of cold, dense water therefore sink and flow across the continental shelf to become a huge current called the Antarctic Bottom Water. From here, branches spread out, moving cold water into the major ocean basins and affecting the rest of the world. The cold water then sinks and travels southwards back to the Indian Ocean and on to the Pacific. Here the current rises, pushed by the flow of water from behind; the water is warmed and so travels near the surface back towards the Atlantic. The whole belt is self-sustaining and could equally work in reverse to the benefit of the north-west Pacific. Convectional movements in the oceans, unlike on the land, can take place both sideways and downwards.

Warm currents can in turn warm the air above and also encourage evaporation and the release of latent heat into the atmosphere. It follows, therefore, that the distribution of heat above the oceans is more efficient than over land areas or within the ground. Land areas tend to warm and cool rapidly due to their efficiency at absorbing insolation. This is referred to as **thermal inertia**. In contrast, oceans and seas have a low thermal inertia and consequently warm and cool slowly.

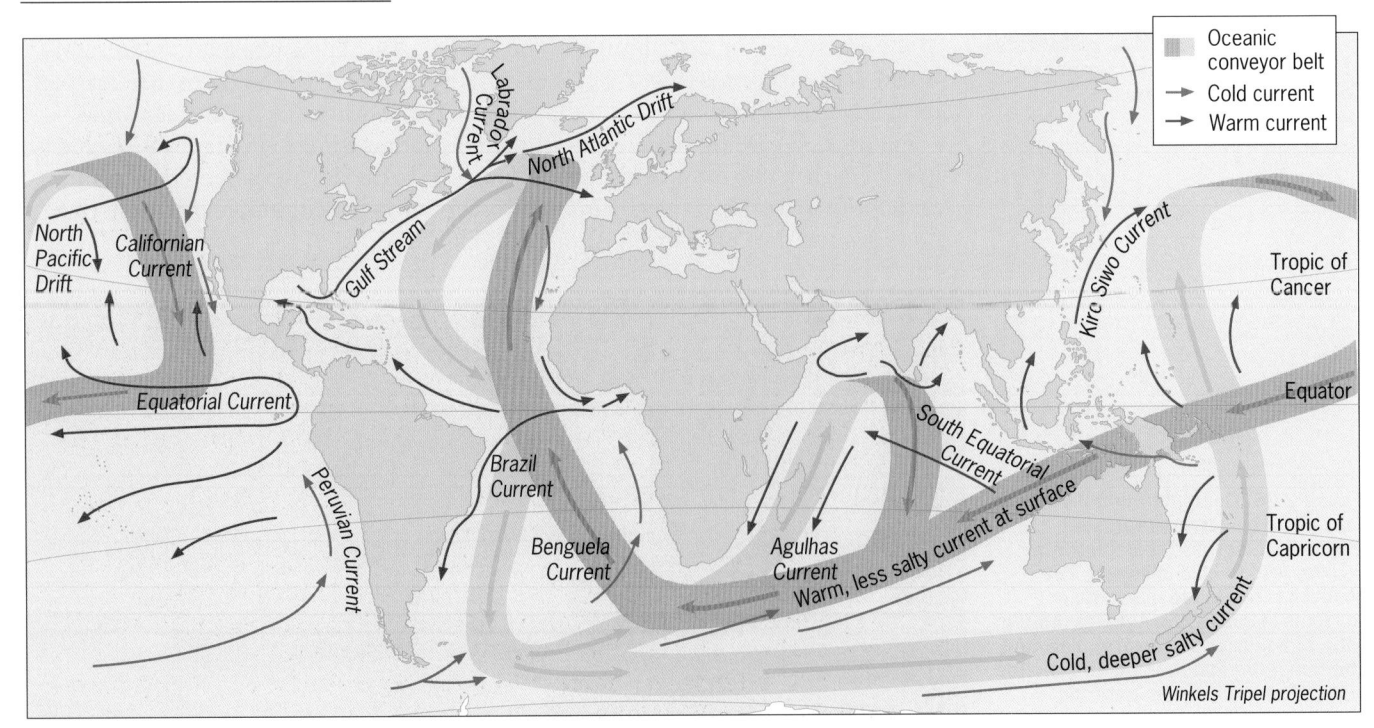

Figure 2.27 Simplified map of the world's major ocean currents showing the oceanic conveyor belt

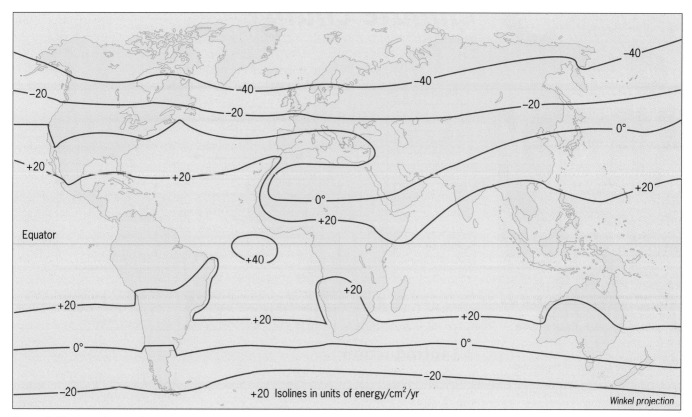

+20 Isolines in units of energy/cm²/yr

Winkel projection

Figure 2.28 Annual net radiation balance of the surface–atmosphere system

It is important to recognise that these spatial variations in energy, and the mechanisms by which energy is transferred, have a fundamental effect on the world's climatic regions and weather. These relationships will be examined more deeply in later chapters.

Summary

- The sun is the ultimate source of virtually all the world's energy. What happens to the sun's energy as it passes through the atmosphere has many implications for life on earth.

- Solar energy is an important natural source of power. Geographical research into the spatial and temporal availability of solar energy is essential if its potential as an alternative energy source is to be fully developed.

- The atmosphere can be divided into a number of layers. Each layer has distinct characteristics that partly result from solar energy passing through them.

- There is a sharp pressure gradient in the atmosphere that decreases with altitude. Temperatures rise or fall with height depending on the atmospheric layer.

- The greenhouse effect is a naturally occurring phenomenon. It results from the way in which certain gases absorb radiant energy from the earth's surface.

- There are important seasonal and diurnal variations in the amount of energy received by the atmosphere and at the earth's surface.

- Some parts of the world experience pronounced variations in their annual radiation budget. This greatly affects the lives of the people living in these countries.

- As the earth's temperatures remain generally constant, insolation received must be balanced by outgoing heat. This is referred to as the solar budget.

- Horizontal and vertical movements of energy are responsible for redistributing energy from areas of surplus to areas of deficit.

3 Climate change

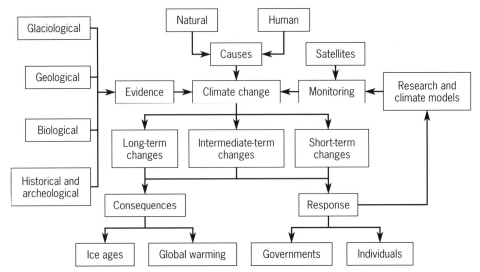

Figure 3.1 Retrieving the carcass of a baby mammoth preserved by permafrost, Susudan, Russia, 1989

Table 3.1 The geological timescale

Period (millions of years Before Present)	Characteristics
Quaternary	
Holocene (10 000 years BP)	Early cities
Pleistocene	Ice ages/ interglacials
Tertiary	
Pliocene (7)	First humans
Miocene (26)	Himalayas form
Oligocene (38)	Alps form
Eocene (54)	First rodents, horses etc.
Paleocene (65)	Extinction of dinosaurs
Cretaceous (136)	Break-up of Pangaea First flowering plants
Jurassic (190)	First birds, mammals
Triassic (225)	Sea levels fall
Permian (280)	Rise of reptiles
Carboniferous (355)	Britain at equator, reptiles
Devonian (395)	Age of amphibians, first fish
Silurian (440)	Plants move to land
Ordovician (500)	Life (shellfish, plants) in sea
Cambrian (570)	Sea covers continents
Proterozoic (2500)	First multicellular organisms
Archaean (4600)	Formation of tectonic plates

3.1 Introduction

Climate change has been widely discussed in recent years. The media have particularly focused our attention on issues such as **global warming**, and the destruction of the **ozone layer**.

We have long been aware of major long-term climatic changes on a scale of thousands or millions of years (Fig. 3.1). The clearest example is the Pleistocene ice age when a cooling of the earth's climate occurred (Table 3.1). Despite such changes, scientists have traditionally viewed the earth's climate as very stable with only minor, short-term (spanning a few years) fluctuations. However, we now know from past geological and biological sources, as well as historical records, that the climate of the earth is constantly changing both in the long and short term. The evidence for such climate change and the nature of the causes raise many issues. We also need to consider if any steps can be taken to influence climate change or reduce its effects.

3.2 Long, intermediate and short-term trends

The atmosphere is constantly changing, so it is necessary at the outset to make a distinction between genuine climate changes and day to day fluctuations in the **weather** (see definitions of weather and climate in section 1.1). We will concentrate on long-term and permanent trends rather than temporary variations of a more random nature. Table 3.2 is an attempt to classify the temporal and spatial scales and is the approach that we will use. The matter of scales can fundamentally affect our perception of climate change. For example, a cooling of the northern hemisphere was noted between 1940 and 1970 and scientists thought this to be the beginning of a long-term global trend. We can now see that this was only a temporary fluctuation – probably linked to a number of volcanic eruptions including Mt Agung in Indonesia in 1963.

Table 3.2 Scales of climate change

Time scale	Long term: billions of years	Intermediate: few million years	Short term: thousands, or hundreds of years
Spatial scale	Global	Global and continental	Global, continental and regional/local

Note: Climate changes at larger spatial scales will have effects at small scales

Figure 3.2 Cavedale, Derbyshire, UK

A further question for us to consider is: What aspects of the climate are changing? The starting point must be temperatures because these affect other atmospheric processes (see section 6.3).

Long-term changes

On a time scale of billions of years there has been little change in global climate. Geological evidence (see section 3.3) suggests that average global temperatures over the last 3 billion years were about 15°C, and over the last million years about 22°C. The range between the annual averages is not likely to have exceeded 4°C. The analysis of ice cores from Greenland and Antarctica and isotope analysis (see section 3.3) indicate that there have been no significant differences in the climate of the two hemispheres.

However, there is a great deal of *palaeoclimatic* (study of past climates) evidence that appears to contradict these conclusions. At the entrance to Cavedale (Fig. 3.2) near Castleton in Derbyshire, the Carboniferous limestone contains coral fragments. Coral reefs, though, are today largely restricted to warm seas in and near the tropics: areas like north-eastern Australia and the Red Sea. Similarly, sandstones in Surrey show evidence of dune bedding associated with the sand dunes found in desert areas. Such examples do not necessarily mean that these latitudes were once experiencing very different climatic conditions. We have learnt that these examples can result from plate tectonics. During the Carboniferous period (see Table 3.1), about 280–355 million years ago, what we now call the British Isles would have been at a more southerly latitude experiencing a tropical climate. Since then, the British Isles have moved northwards to their present position.

Some scientists believe that if it were not for human interference and global warming, the earth would be approaching a cooler period and possibly another ice age. This is partly based on evidence that each ice age or *glacial* phase lasts for about 100 000 years and *interglacials* (warmer phases between two glacial periods) for 15 000 years.

Intermediate-term changes

Periods spanning a few million and thousands of years reveal more impressive climate changes and include the most familiar example of ice ages (Fig 3.3). There have been about four glacial ages during the last one thousand million years, although plate movements, weathering and erosion have removed much of the evidence. We will therefore concentrate on the Pleistocene ice age from 2 million to 14 thousand years ago. The cooling of the earth's climate appears to have started in the Tertiary period about 20 million years BP (before present) (Table 3.1).

Short-term changes

On a time scale of thousands or hundreds of years (essentially since the end of the

Figure 3.3 Temperature change over the last one million years (*Source:* DoE, 1989)

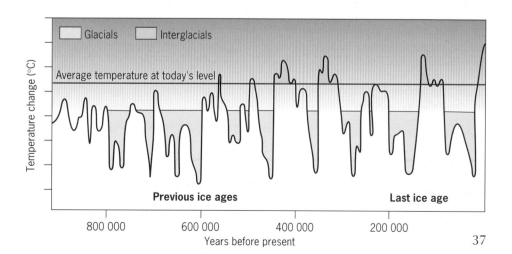

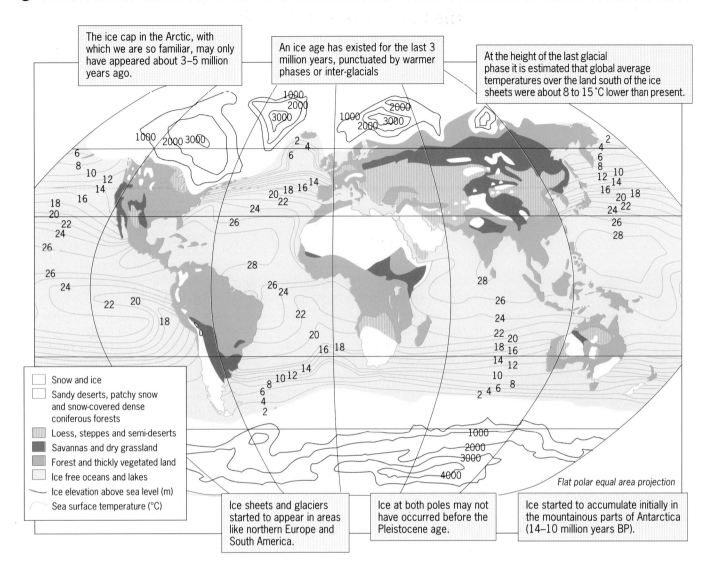

The ice cap in the Arctic, with which we are so familiar, may only have appeared about 3–5 million years ago.

An ice age has existed for the last 3 million years, punctuated by warmer phases or inter-glacials

At the height of the last glacial phase it is estimated that global average temperatures over the land south of the ice sheets were about 8 to 15 °C lower than present.

☐ Snow and ice
☐ Sandy deserts, patchy snow and snow-covered dense coniferous forests
▦ Loess, steppes and semi-deserts
▦ Savannas and dry grassland
▦ Forest and thickly vegetated land
☐ Ice free oceans and lakes
⌒ Ice elevation above sea level (m)
⌒ Sea surface temperature (°C)

Flat polar equal area projection

Ice sheets and glaciers started to appear in areas like northern Europe and South America.

Ice at both poles may not have occurred before the Pleistocene age.

Ice started to accumulate initially in the mountainous parts of Antarctica (14–10 million years BP).

Figure 3.4 The effects of the ice age on the world's natural regions, global sea level and the shapes and sizes of continents (*Source: Cambridge Encyclopaedia of Science*)

?

2a Study Figure 3.4 carefully. Give reasons for the location of the ice caps that developed in South America and New Zealand. An atlas might help you.

b Describe and explain the effects that the Pleistocene ice age had on the parts of the world which were not glaciated.

c Why do you think the desert areas were so much smaller than today?

ice age), climate changes are less impressive although none the less interesting (Fig. 3.5). Although north-west Europe experienced a cooler and wetter time about 2500 years ago, the post glacial period has been marked by short, cold and warmer phases lasting about 100–300 years (Fig. 3.6). It is important to note that much of the evidence for such changes is made from unreliable written sources. For the last thousand years, though, the climate of Britain has remained broadly the same with year-to-year seasonal variations in temperature being as small as 1°C.

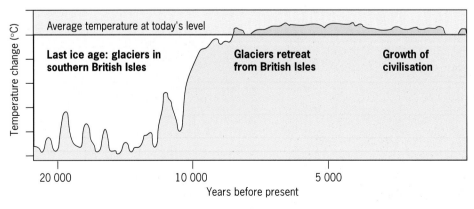

Figure 3.5 Temperature change over the last 20 000 years (*Source: DoE, 1989*)

3a Study Figure 3.5. Estimate the fall in global average temperatures during the glacials.

b Estimate the increase in global average temperatures during the warmer inter-glacials (NB: these values vary considerably in different parts of the world).

4 Using Figure 3.6, estimate the date of the etching in Figure 3.7.

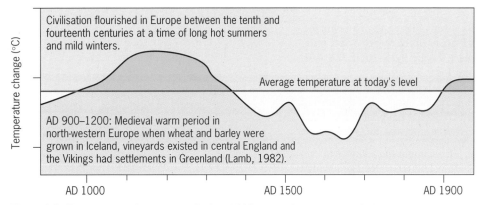

Civilisation flourished in Europe between the tenth and fourteenth centuries at a time of long hot summers and mild winters.

Average temperature at today's level

AD 900–1200: Medieval warm period in north-western Europe when wheat and barley were grown in Iceland, vineyards existed in central England and the Vikings had settlements in Greenland (Lamb, 1982).

Temperature change (°C)

AD 1000   AD 1500   AD 1900

Figure 3.6 Temperature change over the last 1000 years (*Source:* DoE, 1989)

Figure 3.7 Skating on the River Thames at Richmond, UK

3.3 The evidence for climate change

The observation and recording of the world's weather **elements** have only been carried out over the last 150 years. This has largely been restricted to developed areas like Europe and North America but includes some tropical countries. Many parts of the world, particularly in the southern hemisphere and remote regions like oceans and Antarctica, have very short records. In addition, historical sources can provide useful information about climatic extremes (see Table 3.3).

Information about ancient climates has to be obtained using a variety of evidence that has been influenced in some way by past climates. Much of this work is done through the field of *palaeontology* (the study of life in the geological past).

Glaciological evidence

Glaciers

The advance or retreat of glaciers is a clear response to variations in climate. The changes in the three main glaciers near Chamonix in the French Alps, the Mer de Glace (Fig. 3.8), d'Argentierre and Des Bossons, have been extensively studied. Figure 3.9 shows the changes that have occurred since 1644 in the position of the snout of the Mer de Glace. Climate records from Annecy about 60 km west of Chamonix showed a slight temperature rise in each season after the 1860s. The

Figure 3.8 The Mer de Glace, France

5 Study Figure 3.9. Three major advances of the Mer de Glace took place during 1575–1650, 1710–80 and 1816–55.
a To what climatic trend can you link these changes? See section 3.2.
b What has happened to the Mer de Glace since the 1860s?

6 How might the advance or melting and retreat of glaciers in an area like the French Alps have affected people living there?

7 What historical evidence might you look for to assess the changing position of the glaciers near Chamonix over the past 500 years?

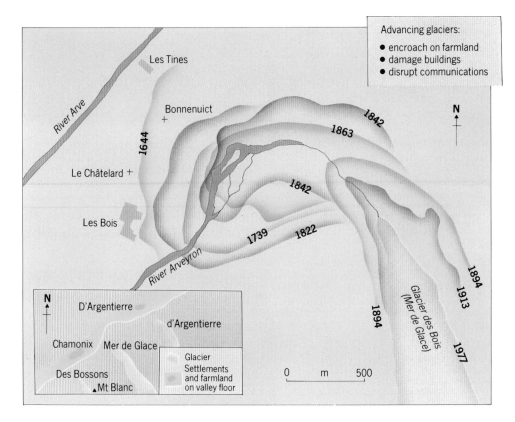

Advancing glaciers:
● encroach on farmland
● damage buildings
● disrupt communications

Figure 3.9 Changes in the position of the Mer de Glace

Figure 3.10 Glaciologist cuts a section of an ice core for analysis of chemical and radiochemical trace elements. Scientists extract ice cores from the highest parts of Greenland and Antarctica where there is very little annual ice melt. Over thousands of years the annual accumulation of snow has gradually compressed to form ice. The cores are divided into layers that represent one year's accumulation.

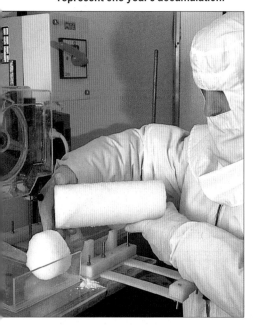

data then suggest that there was a lag of about 10 years in the response of the glaciers. It seems that a rise in temperatures of only 0.5°–1°C over a 60–70 year period is enough to cause glacial retreat. Although some alpine glaciers have been advancing slightly since the 1960s, trends are not clear enough to deduce any significant climatic changes.

Most glaciers in the northern hemisphere have shown a decrease in size in the twentieth century. In fact, many glaciers in the European Alps have decreased by as much as a third. Perhaps this is a reflection of a general rise in temperatures. However, there are exceptions, like the Hubbard Glacier in Alaska that surged in 1986 advancing at a rate of 12 m a day. This suggests that glaciers also respond to local climatic conditions as well as reflecting wider spatial changes.

Ice cores

In 1993 the US Greenland Ice Sheet Project II (GISP2) completed drilling through the ice cap. This was carried out in parallel with a European project, GRIP (Greenland Ice Core Project). These are the latest of a number of attempts to date accurately the shifts from glacial to interglacial periods by using ice cores (Fig. 3.10).

Recent analysis of ice cores spanning the present to 18 000 years ago revealed that Greenland's climate has changed from glacial to interglacial conditions very rapidly in the space of 3–5 years. Glaciologists believe that the melting and collapse of the ice sheets led to a film of water at the surface in the North Atlantic with a very low salinity. This layer has a low thermal heat capacity which leads to rapid heating in the summer, even at these polar latitudes.

Microscopic air bubbles in ice cores can also reveal information about past atmospheric pressure. The density of the air suggests the height of the ice surface when the particular layer of snow fell. Sophisticated sensors can also be used to detect chemical and microparticle changes. Thus it is possible to measure amounts of gases like carbon dioxide (CO_2) and nitrous oxide. The quantity of acid aerosols (fine particles suspended in the atmosphere) can be used to estimate the frequency of past volcanic eruptions. Volcanoes emit sulphur and other particles that have a cooling effect on the climate (see section 3.4).

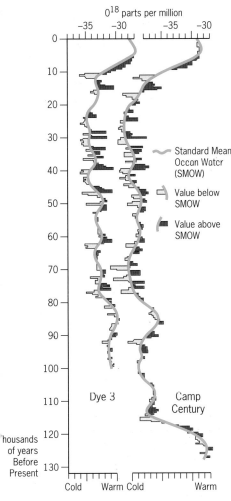

O^{18} parts per million

Standard Mean Ocean Water (SMOW)

Value below SMOW

Value above SMOW

Dye 3

Camp Century

thousands of years Before Present

Cold   Warm   Cold   Warm

Figure 3.11 Profiles of ^{18}O measured along the Dye 3 and Camp Century ice cores from Greenland. (*Source*: Oeschger and Stauffer, 1986)

Figure 3.12 Landsat image of dendritic drainage patterns, Nile Valley, northern Africa

?

8 Divide the curves in Figure 3.11 into three sections, one glacial and two inter-glacials, based on the changing levels of ^{18}O. Identify approximate dates separating these periods.

9 Study Figure 3.12. Identify the patterns labelled **X** and suggest how they reflect climate change.

10 If a fossil was found containing a quarter of the C^{14} in a live specimen today, how old would it be?

Isotope analysis

Isotope analysis has been widely used to complement other techniques to study *palaeoclimatology* (the study of past climates). An isotope is one or more forms of an element differing from each other in atomic mass. The technique involves studying the ratio between two oxygen isotopes, ^{18}O and ^{16}O. In sea water, for example, this ratio is about 0.0021:1. When sea water evaporates during warmer conditions there is an excess of the heavier ^{18}O isotope as the lighter ^{16}O isotope is more easily evaporated. Condensation of the heavier ^{18}O also occurs more readily in the atmosphere. Ice cores therefore reveal a ratio in favour of the heavier ^{18}O isotopes that fell in precipitation during colder glacial periods in the past.

Figure 3.11 shows the results from two ice cores extracted in Greenland. A number of studies have also shown how levels of CO_2, nitrous oxide and other gases as well as dust concentrations mirror the changes in ^{18}O.

Geological evidence

Geological information is widely used to indicate past climate changes in many parts of the world. For example, the extent and size of ancient drainage patterns and fluvial deposits in parts of north Africa cannot be explained by present climates. They reflect past wetter periods called *pluvials* that affected north Africa during the Pleistocene ice age. There are two possible explanations for the greater amounts of precipitation: (a) the movement southwards of the precipitation which normally falls over the Mediterranean áreas in the winter, and (b) the movement northwards of the **Inter-Tropical Convergence Zone** (or **Inter-Tropical Discontinuity** – see section 6.3) during a warmer interglacial.

Geologists use *radio carbon (C^{14}) dating* to find the age of fossils, rocks and sediment. Radiogenic isotopes are formed from the decay of a radioactive parent element. If the decay follows a particular pattern, this can be used as a dating technique. Every radioactive isotope has a known rate of decay or half-life. This is the time taken for half of the isotope to decay.

C^{14}, for example, is a radioactive isotope of carbon that is taken in by plants and animals. It has a half-life of 5370 years so only half of the C^{14} in a live specimen will exist in one that is 5370 years old. C^{14} can be used to accurately date material less than about 30 000–40 000 years old. Fossil water found in Nubian sandstone in north-west Egypt has been dated in this way to 25 000–35 000 years ago, suggesting that a past pluvial existed at that time.

Figure 3.13 Cross section through the trunk of a hardwood oak. The concentric rings correspond to annual growth, indicating the age of the tree. The cracks are due to drying.

Sediment cores extracted from lake beds and the ocean floor can represent a virtually continuous sequence of sediment accumulation. Recent studies of oxygen isotopes in deep-sea sediment reveal oscillations in the earth's climate between warmer and colder periods. The periodicity is 20 000, 40 000 and 100 000 years, related to changes in the position of the earth on its axis in its orbit around the sun (Gribbin, 1988). The biological material in sediment cores is also of great value and can provide a key to understanding changes in the environment that affected life. Many studies have been made of foraminifera, a type of near-surface plankton. Foraminifera, along with other types of plankton, diatoms and radiolaria, are very sensitive to water temperature changes. Their presence in, or absence from, sediment indicates fluctuations in the climate. Radio carbon dating of marine organic matter can be used to date the beginning and end of warmer periods.

We have to take care, though, when studying geological evidence and in reaching conclusions about past climatic changes. There are a number of reasons for this:

- There are often gaps in the spatial records of past climates in soils or features.
- In many parts of the world sediments and soils have been disturbed by human activities, particularly agriculture.
- Later ice advances often removed or altered much evidence from previous phases of glaciation.
- A lack of organic material may make the dating of past events difficult.
- Many variables are involved in relation to past vegetation and other forms of life. These are discussed more fully in the following section on biological evidence.

Biological evidence

Plant growth is strongly related to climate, particularly temperatures, sunlight and precipitation. Scientists have developed a number of techniques which link the features of plants to past climates.

Dendrochronology

Dendrochronology originates from two Greek words: *dendron*, meaning 'tree', and *chronos*, meaning 'time'. The technique involves the study of the annual growth rings of trees (Fig. 3.13) that reflect the climate of an area. In semi-arid environments, the availability of moisture affects the amount of annual growth and hence ring width. In areas of boreal (coniferous) forest, such as northern Canada, the length of the growing season determined by temperatures and frost is the main control on annual cycles of growth. As a tree gets older, each year's growth is spread over a wider circumference. Mathematical techniques can now compensate for this and be used to analyse annual growth rings in relation to climate. For example, the studies of the bristle cone pine trees of the south-west USA are now well known. These have enabled the construction of climate records covering the last 8000 years for this part of North America.

Pollen analysis

The characteristics of plants, including rates of growth and flowering and the production of pollen, are influenced by climate. Temperature, precipitation, the gases in the atmosphere and length of the growing season are particularly important. As climate changes, so too will the plant communities and species in an area.

In the early 20th century, Scandinavian scientists realised that the pollen preserved in peat and sediment could be used to reconstruct past climatic conditions. Sediment cores are particularly useful because, as the layers vary in thickness and content, they reflect how the environment was changing. For example, a core from a lake may contain layers with a high dust content. This may be wind blown material from a periglacial environment. The varying thickness of

?

11 Study Figure 3.15.
a Describe the vegetation from the end of the last ice age to the end of the pine age.
b How did the tree species change after the pine age?
c How do these changes in tree and plant species relate to climate change?

sediment layers may correspond to variations in precipitation – greater amounts of sediment being deposited in particularly wet years. Radio carbon dating can then be used to date layers and therefore place any pollen into an age context.

One flower from a plant can produce 10 000–1 000 000 pollen grains and each can be as small as 10–100 µm (Fig. 3.14). Only small samples of sediment are therefore needed for analysis. The frequency of different types of pollen indicates which species were dominant at that time, so reconstructing the climate simply involves knowing what climatic conditions are favoured by those species today. An application of this technique shows how forest vegetation returned to the British Isles at the end of the ice age (Fig. 3.15).

Historical and archeological evidence

Historical evidence includes a variety of sources (Table 3.3). These are, though, mostly restricted to the last 800 or 900 years and to only a few areas of the developed world. In addition, reliable meteorological observations date only from the mid-nineteenth century.

During the 10th–14th centuries, a number of civilisations in the Americas changed dramatically. Their study can be used to illustrate the value of archeological evidence. Tiwanaku, the highest urban centre in the new world, collapsed

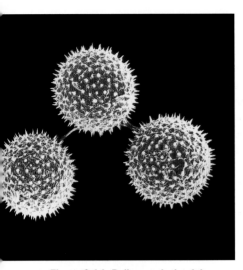

Figure 3.14 Pollen analysis of the common mallow: microscope slides are prepared and the individual pollen grains then identified and counted

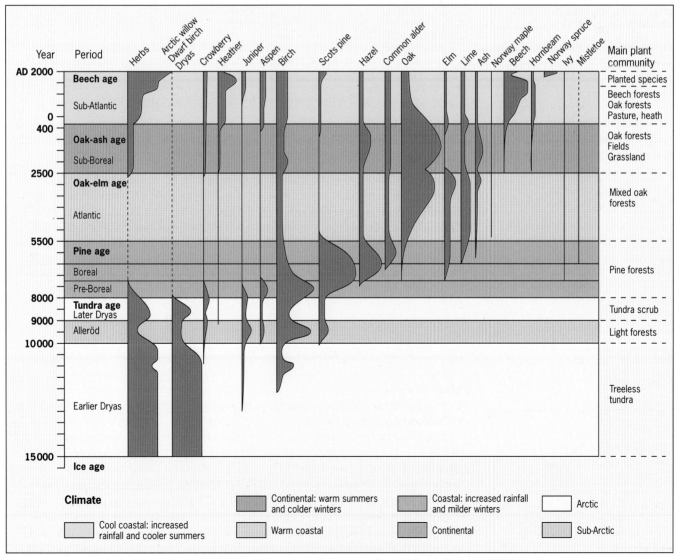

Figure 3.15 Development of British forests from prehistory to modern times (*After*: Vedel and Lange, 1960)

43

Table 3.3 Historical sources of climate change

Printed material: weather diaries and records kept by amateur meteorologists

Paintings and engravings: cave paintings in France show mammoths that lived during the ice age; Egyptian wall paintings show animals like lions and elephants that survived in northern Africa when the climate was wetter than today; a number of famous artists, e.g. Bruegel, painted landscapes during the 'Little Ice Age' showing conditions far more extreme than today

Photographic material: photographs can record severe events like blizzards, floods or storm damage very accurately

12 G. Manley built up a picture of climate change in Britain over the last 1000 years. Use a library to see what you can find out about Manley's work on historical sources and climate change. Write a short summary of your findings.

13 What problems might exist in drawing conclusions about past climate change from sources like diaries and personal observations?

14 Essay: Critically evaluate the evidence for climate change.

during AD 1000 and 1100. There was also a marked decline in the Anasazi cliff dwellers in the American south-west around AD 1300. Archeologists have noted how these cliff dwellers used timber from the ponderosa pine and spruce trees, although the climate of this area is now too dry to support such species. Researchers have attributed these changes to the climate change of the medieval period, in particular to fluctuations in rainfall and the consequent effects on vegetation and agriculture.

3.4 Natural causes and effects of climate change

Solar activity
One obvious cause of climate change might be changes in the amount of energy emitted by the sun. Measurements from spacecraft suggest that this varies rhythmically; scientists have identified 11, 22, 100 and 79/80-year cycles. The 100 and the 79-year cycles can be combined to produce a 179-year cycle that scientists relate to patterns detected in tree ring studies.

Earth geometry and Milankovitch cycles
The earth's geometry also varies, which in turn affects how much solar energy different parts of the earth receive. In the 1920s a Yugoslav scientist, Milutin Milankovitch, calculted three ways in which the earth's geometry varies. These are known as the **Milankovitch cycles**:

1 The earth's orbit around the sun changes from near-circular to being more elliptical (oval) on a cycle of about a million years.

2 The earth rolls or 'wobbles' slightly on its axis on a cycle of 40 000 years.

3 The axis of the earth moves round slowly on a cycle of 20 000 years, affecting the distance of different areas from the sun on midsummer's day.

The Milankovitch cycles have been matched quite successfully to the occurrence of ice ages dated by other methods (see section 3.3). However, the existence of periods in the earth's geological history of as long as 250 million years without glaciation would suggest that other factors are involved in climate change. The use of Milankovitch cycles for defining change has also been criticised because each cycle would barely influence the amounts of solar energy reaching the earth. They mainly change the *distribution* of energy over the earth's surface. They therefore have a large effect on *seasonality*, without changing the actual *amount* of energy within the system.

Plate tectonics
Plate movements can be responsible for moving land areas into different climatic regions. However, can plate movements actually bring about climate change? There is some evidence that the existence of a land mass at the poles, or surrounding the poles, encourages ice sheets to develop. These land areas may prevent warm ocean currents from reaching the polar areas, so snowfall is less inclined to melt and can accumulate to form ice sheets.

It is fascinating to consider the effects that the formation of the Himalayas and the Tibetan plateau has had on global climate change. Research suggests that a general cooling of the world's climate resulted. Some global climate models show that, without the Himalayas and the Tibetan plateau, the Indian monsoon would not occur (see section 10.4). For almost 250 million years the earth's climate was warm and wet, then about 40 million years ago a marked cooling began. This coincided with major plate movements, including the merging of India with Asia that led to the uplift of the Tibetan plateau. There is no doubt that plate movements and the way they shape the earth's crust do influence climate change.

15a Copy Figure 3.16. Assume that there is a build up of snow and ice cover on the earth. Place the following labels into the boxes in the correct sequence to show how increasing albedo translates the input into an output.
- Output: Energy reflected from snow and ice
- Snow and ice cover increases
- Input: Short-wave solar energy is reflected
- More energy is reflected and radiated
- Melting is reduced

b Add positive or negative signs beside each arrow to indicate the relationships between each component in the system.

c Does the diagram you have drawn provide an example of positive or negative feedback?

16 Attempt to construct a similar diagram for a disturbance to the system involving a reduction in snow and ice cover. Does this illustrate positive or negative feedback?

Ice-albedo effect

Snow and ice form an integral part of the climate system. Both can reflect huge amounts of incoming solar energy. This ability of a surface to reflect solar radiation is known as **albedo** (see section 2.4). Snow and ice have an albedo of about 80 per cent, compared with 25 per cent for grass. In consequence, there is an important feedback relationship between the earth's climate and snow/ice cover. A build-up of snow and ice cover has a cooling effect on the earth's climate. In contrast, a reduction in cover leads to more heat being absorbed by the ground and contributes to a warmer atmosphere. Obviously many other factors, like solar activity, affect the world's climate which can reverse these trends.

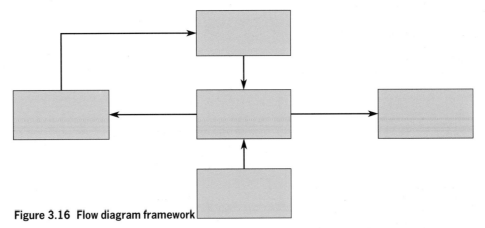

Figure 3.16 Flow diagram framework

Oceans and the El Niño current

The oceans and seas of the world have their own circulations of warm and cold water. In many ways these are like the weather and climate of the atmosphere. For example, currents like the Gulf Stream flow year after year just like the Trade winds (see section 6.6). There are huge eddies in the oceans over 60 km across which can travel for hundreds of kilometres before fading. The undersea circulations are also closely linked with the atmospheric circulations (see section 6.6).

One clear example of the links between oceanic movements and climate change concerns El Niño events. These are related atmospheric/oceanic phenomena which disrupt the climate across much of the rest of the world from East Africa to the shores of the Americas (Fig. 3.17). **El Niño Southern Oscillations (ENSOs)** occur every two to nine years. High sea surface temperatures in the western Pacific trigger a reversal in the normal westward flow of the Trade winds and ocean currents that flow across the tropical Pacific from the Americas towards Asia. El Niño, meaning the boy child, is so called because its waters reach the coast of the Americas at about Christmas.

Many climatologists have produced computer forecasting models, largely based on sea temperatures, to try to predict ENSOs and their effects worldwide. One of the most successful is at the Lamont–Doherty Geological Observatory at Columbia University in New York. Computer models forecasted the ENSOs of 1982–3 (particularly strong), 1986–7, 1989–90 and 1991, in some cases up to eight months ahead. However, ENSOs are turning out to be less predictable than expected. ENSOs generally last for about 18 months, although the 1990 ENSO reached its peak during early 1992, weakened suddenly in mid-1992 and then strengthened again in November 1992.

The devastating consequences of ENSOs are proving even harder to predict. Researchers' efforts are nevertheless important as people can take steps to manage the worst effects of climate changes. In densely populated tropical areas, warnings of severe weather could save thousands of lives. Some scientists are advising that global warming would intensify the build-up of warm water and that increased sea temperatures could then lead to more intense El Niño events.

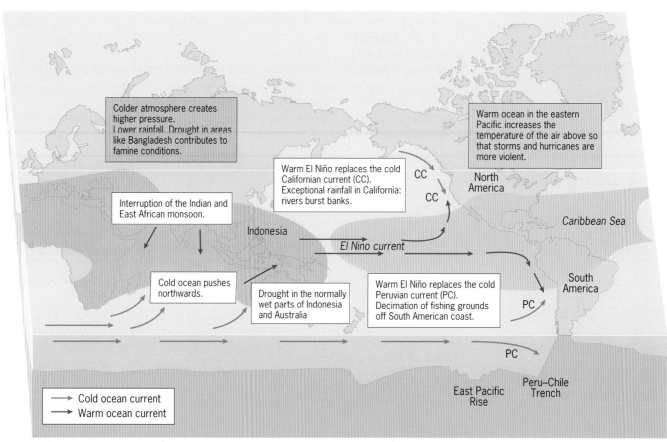

Colder atmosphere creates higher pressure. Lower rainfall. Drought in areas like Bangladesh contributes to famine conditions.

Warm ocean in the eastern Pacific increases the temperature of the air above so that storms and hurricanes are more violent.

Warm El Niño replaces the cold Californian current (CC). Exceptional rainfall in California: rivers burst banks.

Interruption of the Indian and East African monsoon.

Cold ocean pushes northwards.

Drought in the normally wet parts of Indonesia and Australia

Warm El Niño replaces the cold Peruvian current (PC). Decimation of fishing grounds off South American coast.

Indonesia

El Niño current

CC

CC

North America

Caribbean Sea

South America

PC

PC

East Pacific Rise

Peru–Chile Trench

→ Cold ocean current
→ Warm ocean current

Figure 3.17 The impact of an El Niño oscillation

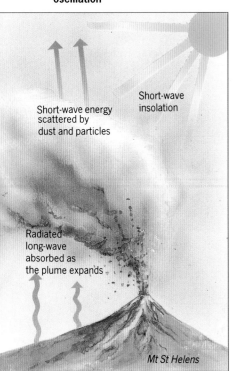

Short-wave insolation

Short-wave energy scattered by dust and particles

Radiated long-wave absorbed as the plume expands

Mt St Helens

Figure 3.18 Atmospheric effects of the eruption of Mt St Helens, Washington, USA

Volcanic activity

There have been a number of volcanic eruptions since about 1980 which illustrate clearly the effects that volcanic eruptions can have on climate change. When Mt St Helens erupted on 18 May 1980, it sent huge quantities of ash and sulphate aerosol (finely suspended) particles high into the atmosphere with a number of atmospheric effects (Fig. 3.18). The most significant result, though, was the alteration of the earth's albedo (the reflectivity of a surface) and an increase in reflected energy. Satellite readings the day after the eruption showed albedo values about twice normal levels over the states of Wyoming and Montana. However, the blast from Mt St Helens was mainly lateral and the ash settled rapidly. Because little material was sent into the **stratosphere**, there was little climatic disturbance. Computer estimates showed the effect of the eruption on annual average temperatures for the northern hemisphere was less than 0.1°C. For an eruption to have a substantial effect on climate, particles need to be carried higher into the atmosphere.

In June 1991, the largest volcanic eruption for 80 years occurred in the Philippines. The eruption of Mt Pinatubo was estimated to be of a magnitude about ten times greater than Mt St Helens. According to the US Geological Survey, approximately 5–8 km³ of ash and aerosol particles were ejected into the atmosphere, with the cloud extending to a height of 24 km into the stratosphere. In the following weeks, satellites observed particles from the eruption encircling the globe (Fig. 3.19). Satellite temperature measurements confirmed that the dust was shading part of the earth. It has been estimated that, as a result, average global temperatures could be reduced by over 1°C for up to 5 years. The eruption also had consequences for ozone depletion (see section 5.3) and may have implications for global warming (see section 4.3).

Principally, volcanic eruptions seem to have a cooling effect on the earth's climate and may be a contributory cause of glacial phases. In the early twentieth century the relative lack of eruptions may be associated with slightly higher average global temperatures. The full cause and effect link between volcanic eruptions and climate change is still uncertain, though. Other variables are involved which may counteract the cooling effect. Some scientists believe that the El Niño effect in 1991, which caused a warming of the atmosphere, could compensate for the cooling brought about by Mt Pinatubo. This remains an area for research.

3.5 Climate change: human causes, effects and response

Changes in vegetation

We have already seen that there are close links between vegetation and climate. Variations in the spatial distribution of different plant species may be one indicator of global warming.

The impact of deforestation of the world's rainforests has been widely publicised and researched. Figure 3.20 illustrates the natural balance that exists between the vegetation and the atmosphere. However, destruction of the rainforests can seriously disturb this balance, leading to positive feedback that results in semi-arid conditions and other environmental problems, such as soil erosion.

Similarly, Figure 3.21 shows how clearly areas of forest stand out against the stark whiteness of a snow-covered landscape. What you may not consider is that this has climatic influences. The boreal forests of the northern hemisphere absorb more sunlight during winter months than the adjacent treeless, snow-covered areas. The forests are therefore warmer (Fig. 3.22). Clearly, altering the vegetation cover could have climatic consequences.

Deforestation in northern Europe

Boreal forests consist largely of evergreen needle-leaf tree species such as the Norwegian spruce. In northern Europe they cover an area of approximately $450\,000\ km^2$ (more than $4\,000\,000\ km^2$ in the northern hemisphere). These are of great importance to the northern European countries for their commercial value (Table 3.4).

Figure 3.19 Nimbus-7 satellite image of the sulphur dioxide cloud released after the eruption of Mt Pinatubo, Philippines, 1991. The false colour map was made on 30 June 1991, 18 days after the eruption. The highest concentrations of gas are shown as green, centred over the Atlantic Ocean. By this time the cloud had spread in a belt eastwards over the Pacific Ocean and it completed its circuit of the earth by 5 July. As the cloud moves, much of the gas reacts with atmospheric water to form a sulphuric acid aerosol.

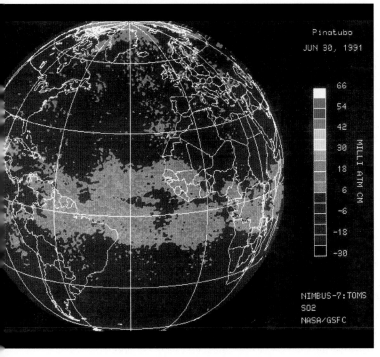

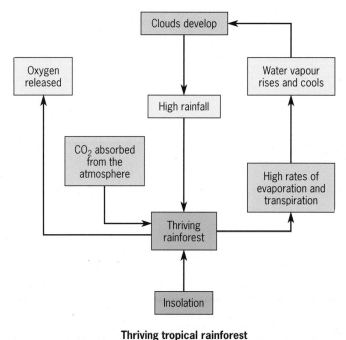

Thriving tropical rainforest

Figure 3.20 Tropical rainforest and the atmosphere system

Figure 3.21 Contrasting snow cover: view of boreal forest and open land, Alaska, USA

?

17 Draw a flow diagram to show the positive feedback in the atmosphere caused by deforestation. Use Figure 3.20 to help.

A computer model developed at the National Centre for Atmospheric Research in Boulder, Colorado, USA, shows how important boreal forests might be to global climate stability. When the computer model considered removing all the forests north of 45°N and leaving bare soil, the result was a dramatic cooling. At 60°N in April, the average temperature fell by 12°C. In late summer too, when there was no snow and the soil absorbed as much sunlight as the forest it replaced, the cooling was still 5°C. The impact of deforestation was felt throughout the year because the winter cooling was so great that it significantly affected ocean temperatures.

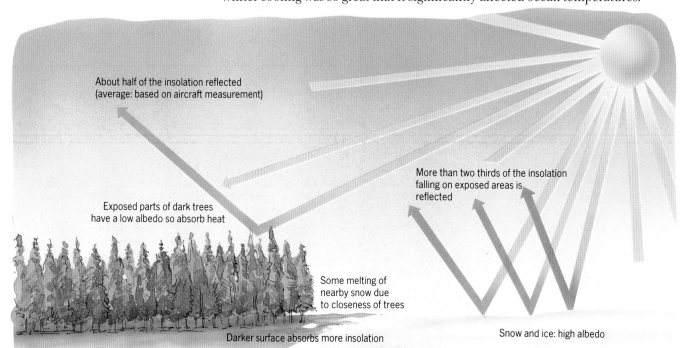

About half of the insolation reflected (average: based on aircraft measurement)

Exposed parts of dark trees have a low albedo so absorb heat

More than two thirds of the insolation falling on exposed areas is reflected

Some melting of nearby snow due to closeness of trees

Darker surface absorbs more insolation

Snow and ice: high albedo

Figure 3.22 The effect of an open snow-covered landscape and boreal forest on incoming solar radiation

Table 3.4 Ranks of top ten countries for softwood products, 1993

Country	Paper and paper board	Wood pulp	Softwood sawnwood	Softwood roundwood
Brazil	10	7	8	6
Canada	3	2	3	3
China	4	–	5	4
CIS	6	4	1	2
France	9	9	10	9
Finland	7	6	9	7
Germany	5	8	7	8
Japan	2	3	4	10
Norway	–	10	–	–
Sweden	8	5	6	5
USA	1	1	2	1

Scandinavian countries highlighted

– not ranked in top ten

18 Use newspapers on a CD ROM, or in a library, for the end of March/early April 1995. Write a summary of the findings and recommendations of the International Climate Conference held in Berlin.

19 Essay: Discuss the extent to which climate change is natural or the result of human activities.

Because oceans warm and cool slowly (**thermal inertia**) the cooling effect lasted through the summer.

Although the results of this computer model may need to be treated with caution, they do, however, suggest the importance of replanting programmes (afforestation). They also raise important questions about how logging and natural fluctuations in these forests may influence climate change.

Pollution

From our examination of the natural causes of climate change there was some indication that the earth's climate should be slowly cooling as another ice age approaches. For about thirty years after World War 2 there was a general, although rather erratic, fall in average global temperatures. Since then, the average has been rising.

The burning of fossil fuels and deforestation have both resulted in a steady rise in CO_2 levels in the atmosphere. There has also been an increase in other gases which, along with CO_2, absorb outgoing radiant energy from the earth. The effect is an overall warming of the atmosphere. Are these changes, though, part of a long-term global trend or the result of human activities? The causes and consequences of global warming are complex and the subject of debate (see Chapter 4).

The response to climate change

International action and co-operation are essential for the solution of problems created by climate change. The efforts of individual countries are important e.g. in reducing output of greenhouse gases, but are of little use if other countries expand their levels of pollution. In 1988 the Intergovernmental Panel on Climate Change was established to encourage research and act as an international forum for ideas and advice (see also section 4.5).

Our ability to adapt to future climate change will partly depend on processing data in a predictive way. *General circulation models*, the carbon cycle model and data on CO_2 levels are crucial for understanding global warming. Much of our discussion assumes that the atmospheric system is governed by laws. This indicates that the more we learn, the better our models and understanding will become. Are there so many variables involved, though, that unexpected, natural fluctuations will always take place? It seems wise to continue with our models yet recognise we may never predict climate change accurately.

Summary

- Climate change has been a widely discussed aspect of environmental change in recent years. This is even to the extent that sometimes minor weather events are inaccurately linked to longer trends.

- Climate change can be examined at a variety of scales. The long-term, covering billions of years, the intermediate scale, spanning millions of years, and the short-term changes that occur over a few thousand years.

- Over billions of years the earth's climate has been relatively stable. Ice ages represent one of the most dramatic examples of climate change at the intermediate scale. There have been several warmer and colder periods during the last 10 000 years since the last glacial period.

- Information about ancient climates has to be obtained by using geological, biological or glaciological evidence that has been influenced in some way by past climates.

- Increasingly, human activities are being linked to climate change. The extent to which these changes are the result of human behaviour or natural processes, like volcanic eruptions, is a major question and one that is not easy to resolve.

- We can respond to climate change in different ways. These include reducing our activities that are harmful to the atmosphere and adapting to a different climate.

- International action and co-operation will also be essential to the solution of problems created by climate change.

4 Global warming

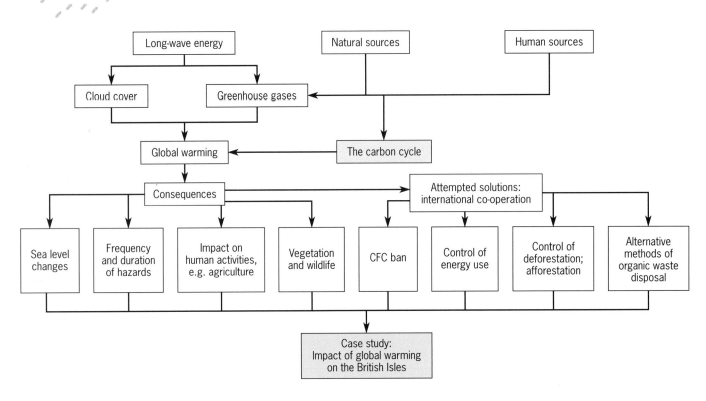

4.1 Introduction

The **greenhouse effect** (see section 2.3) is a naturally occurring phenomenon which people have known about since the mid 19th century. However, it was not until the 1980s that **global warming** was seen as 'a serious threat to the world' (Margaret Thatcher). We are becoming more concerned by this because of an apparent sudden rise in global temperatures, believed to be the result of human activities.

The media have given much attention to global warming (Figs 4.1 and 4.2). Nevertheless, while there has been almost unquestioning acceptance of such theories by many people, others remain sceptical.

Melting ice

There will come a time, perhaps a thousand years hence or even earlier, when large parts of the South will be permanently under water and no barriers will be able to prevent it. Gigantic lakes will cover much of the Home Counties. Coastal towns like Dover and Brighton will be submerged. The Thames will swell until it resembles the Orinoco.

As the millennia go by, increasingly less will be seen of the once prosperous South. It will all be part of the Channel and the North Sea. Only a few islands will remain dotted around the storm-tossed ocean, the tops of the Cotswolds and the tors of Dartmoor. Nuclear submarines will prowl through Aylesbury's High Street, and eels will make their lairs in the crypt of St Paul's.

Figure 4.1 Exaggerated claims about global warming (*Source: Sunday Telegraph*, 11 Dec. 1988)

Figure 4.2 Six-month summers coming to Britain?

The Costa del BRIGHTON!

Weather watchers revealed yesterday that resorts like Bognor and Brighton will enjoy six-month summers as global warming turns them 'semi-Mediterranean'.

1   How do you think we should approach the issue of global warming and react to claims like those being made in Figures 4.1 and 4.2?

2   Study Figure 4.3.
a   In rank order, list the seven warmest years which have taken place since 1850.
b   Do you think this suggests that global warming exists, or not? Give reasons for your answer.

4.2 Does global warming exist?

We saw in Chapter 3 that the world's **climate**, both in the short and long term, is constantly changing, although the fluctuations are relatively small. In this context, we need to consider whether global temperature rises are just part of natural long-term patterns. Ice ages and warmer periods may be examples of the system being disturbed and natural *negative feedback* mechanisms restoring a more stable atmospheric state. Alternatively, global warming may be an example of *positive feedback*, a more rapid and potentially catastrophic disturbance to natural processes caused by human activities.

Figure 4.3 shows how average world temperatures have changed since 1850. The environmental disaster of the dustbowl in the Midwest USA and a series of hot summers in the 1930s were understood as a sign of global warming. You will see that during 1940–70 there was an overall fall in average temperatures, in fact about 0.2°C. There was even talk at this time of another ice age (see section 3.2) and scientists suggested that a number of successive hot summers in the 1980s were random fluctuations. Taking a short period of time from the sequence, it would be easy to argue a case for either warming or cooling.

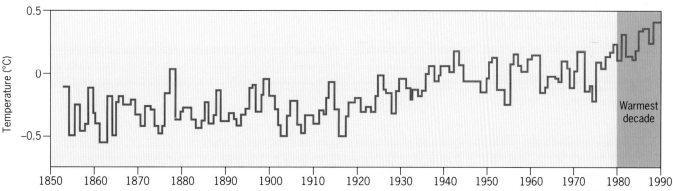

Figure 4.3 Average world temperatures for the last 140 years (*Source:* Met. Office, CRU University of East Anglia). These results are based on very comprehensive data: British researchers use temperature figures from over 1100 stations on every continent and over a million readings have been taken at sea. The data are used to compile a monthly global average of temperatures.

British research findings

In the early 1990s, scientists became more certain of a notable warming of the world's climate. Researchers from the Meteorological Office and the Climatic Research Unit at the University of East Anglia found 1990 to be the warmest year on record since 1850 when reliable records began (Fig. 4.3). The global mean temperature in 1990 was 0.39°C above the average temperature for the period 1951–80.

Although some areas, like the Far East and most of the USA and southern Canada, were notably warmer, others like north-eastern Canada and Greenland were cooler than the long-term average. There are still some spatial gaps, notably around Antarctica, but satellites are increasingly being used to provide data for these areas. Overall, some scientists are therefore preferring to refer to global climate change rather than global warming.

The IPCC model

The Intergovernmental Panel on Climate Change (IPCC), in its first major report in 1990, estimated a global warming of 3°C by AD 2100 and an upper limit of 5°C. This was based on an estimated increase in greenhouse gases (see section 2.3) at the then current rates. The estimate of global warming was later modified in a supplement published by the IPCC which suggested that the global average warming is unlikely to lie ouside the range 1.5–4.5°C with a doubling of atmospheric carbon dioxide (CO_2).

This scientific data formed the basis for the debate on global warming at the Earth Summit in Rio de Janeiro in 1992. There was a great deal of discussion and

USING satellite data, they (Christy and McNider) initially found no global warming trend between 1979 and 1993. They adjusted their results to allow for the effects of two volcanic eruptions and four changes in Pacific circulation – and came up with a warming trend of 0.09° over the period – one third of what the IPCC predictions suggest. It is time for a major re-evaluation they say.

'What we have shown is that when you look at real data, the earth's atmosphere is not warming at the rate previously thought,' says Christy.

'What the new data indicates to me is that there is less pressure to do something drastic in terms of global warming. The earth is simply not warming up to the extent that people have said.

Christy's results have to be taken seriously because they are the first to come from satellite data and are likely to give a more accurate picture of temperature change in the atmosphere than earth-based measurements. In that, they differ from all greenhouse research to date.

Figure 4.4 Christy and McNider's view of global warming (Source: Oliver Morgan, Sunday Express, 27 Mar. 1994)

'. . . the world's leaders foresaw the future and depicted an alarming vision of flood, drought and famine' (Morgan, 1994). Following the Earth Summit, many governments released documents and plans outlining steps to combat global warming; Britain's scheme was launched in January 1994 and one volume was called *Climate Change* (see section 4.5).

Christy and McNider's results

In the same week that *Climate Change* came out, a major challenge to the idea of global warming was published in the scientific journal *Nature*. Two American scientists, John Christy and Richard McNider (University of Alabama), claimed that satellite data showed there was considerably less global warming than had previously been suggested (Fig. 4.4).

The IPCC predictions had been based on mathematical *models* that use ground-based data. They include feedback mechanisms and obscuring factors that influence temperature in the climate system like ocean flows, solar output and the effects of clouds. Christy and McNider's data suggest that starting from ground-based data for the models has led to an over-estimate of global warming. Christy and McNider's research is too good to be ignored but they have been described as 'voices in the wilderness' (Morgan, 1994). Research scientists are now arguing that, although the overall temperature changes are not definite proof of human-made global climate change, there is growing evidence.

J Christy

I think at that point it was a political exercise for many of the people there [the Rio Earth Summit]. When it comes down to it, scientific fact does not enter into the debate. Some people have taken this issue to an extreme. In the media, especially on televeision, they have taken it to extremes that are ridiculous.

Figure 4.5 Attitudes to global warming (Source: Oliver Morgan, Sunday Express, 27 Mar. 1994)

If you start preaching the end of the world being nigh and people don't see anything happening, there is a risk that the public will feel the issue has been over-sold, and then they might well shut their eyes to it.

It seems to me that the scientific jury is still out on the nature and the level of risk from global warming and the greenhouse effect.

I think the public jury isn't out any more. We have moved, in the public domain, to come to judgements ahead of the consensus of the scientific community.

There is a danger. I think we could, in the long-term, undermine public confidence in science and technology, and that I would worry about.

If we devalue scientific expertise by being sold unreliable conclusions as if they were reliable, we might expect some public disillusionment in the long-term, and in my view, that would be bad for both science and public.

Professor J Durant

Figure 4.6 Problems ahead for science and scientists (Source: Oliver Morgan, Sunday Express, 27 Mar. 1994)

?

3   Although Christy and McNider's findings differ from most other research to date, why should they be considered seriously?

4   Explain what Christy suggests are important influences on attitudes to global warming in Figure 4.5.

5   Consider the different research findings relating to global warming and study Figure 4.6. How do you think we should proceed when making environmental decisions that relate to global warming? Give reasons for your views.

4.3 The nature of the atmosphere and global warming

In this section we will examine the hypothesis that global warming results from human disturbance to the atmospheric system (Fig. 4.7). For this, we need to understand global warming in the context of the chemical composition of the atmosphere. This is because certain gases in the atmosphere, and their response to *inputs* of radiated energy from the earth, are responsible for global warming. Remember, the atmosphere is an ocean of gases above the earth's surface approximately 1000 km deep (see Figure 2.9). However, it is impossible to be precise about its depth as, at its outer edge where there is no more than the merest trace of gases, the boundary with space can barely be detected.

The chemistry of the atmosphere, unlike its physics, is relatively simple. There are two main categories of gases, *permanent* and *variable*. Permanent gases are

Figure 4.7 The atmospheric heat transfer system

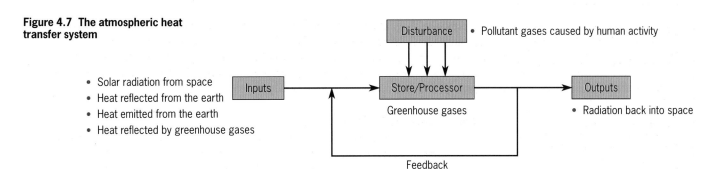

- Solar radiation from space
- Heat reflected from the earth
- Heat emitted from the earth
- Heat reflected by greenhouse gases

6 Look at the information in Table 4.1. Which do you think are the most important greenhouse gases? Give your reasons.

those largely constant in amount and are dominated by nitrogen (75 per cent) and oxygen (23 per cent). There are many others in smaller amounts, such as hydrogen, neon, argon and helium. Variable gases, often called greenhouse gases, are those which vary in amount both spatially and temporally and there are two of major importance, ozone and *water vapour* (Table 4.1). Other variable gases include CO_2 and oxides of nitrogen and sulphur. These are partly natural, being emitted from volcanoes and CO_2 as a waste gas from respiration. However, they are also

Table 4.1 Location and sources of the major greenhouse gases, and their influence on global warming

Greenhouse gas	Where it is found	Where it comes from	Function	Contribution to global warming (%)
Water vapour	Concentrated in lowest 10 km of **troposphere**.	Surface water e.g. lakes, rivers, seas, oceans. Moisture loss from plants and other organisms. Largely constant in amount with spatial variations. Not significantly affected by human activities.	Water vapour only absorbs a small amount of outgoing radiation. Water droplets absorb wave lengths not absorbed by gases/vapour and radiate this heat back to the ground.	Clouds and water vapour account for over 97% of the natural greenhouse effect
Carbon dioxide (CO_2)	Concentrated in lower troposphere.	Released by animals during respiration; burning of fossil fuels and vegetation e.g. rainforests. Annual fluctuations reflect seasonal vegetation growth, especially in northern hemisphere: in summer huge amounts of CO_2 are absorbed, while in winter the process slows. The southern hemisphere does not feature strongly as there is less land area.	Absorbs long-wave radiation from the earth.	about 50
Methane (CH_4)	Concentrated in lower atmosphere.	Released as bacteria break down organic matter: marshes (also known as 'marsh gas'), swamps, tundra, wetlands (used for rice-growing), disposal of organic waste, waste from digestion in mammals. An increase in ruminating animals e.g. sheep, cows has caused levels to rise over the last 200 years.	Absorbs long-wave radiation from the earth.	about 18
Nitrous oxide (N_2O)	Concentrated in lower troposphere.	Combustion of fossil fuels in power stations and transport. Also from denitrifying bacteria which break down nitrate and nitrites – increasing due to use of nitrate fertilisers.	Absorbs long-wave radiation from the earth. Concentrations of N_2O are rising at approx. 0.3% per year.	about 6
CFCs	Ascends to the ozone layer: 15–50km above earth's surface.	Released from aerosol sprays, 1960s–1970s. Used in refrigeration.		24 (Survive about 100 years in the atmosphere. By 2030, it is estimated they will account for 33% of global warming.)
Ozone (O_3)	Largely concentrated 15–50 km above earth's surface. About 10% in the troposphere.	Formed naturally from 3 oxygen atoms due to UV radiation. Tropospheric ozone is human-made resulting from complex chemical reactions between pollutants and sunlight. Sources of these pollutants are nitrogen oxides from power stations and hydrocarbons from cars and other types of transport.	Ozone is vital to life on earth: it filters harmful short-wave UV radiation from the sun.	This gas varies spatially, so contributions are difficult to estimate.

7 Study Figure 4.8. Calculate the total percentage increase of carbon dioxide levels in the atmosphere between 1850 and 1990.

8 Use Figures 4.8 and 4.9.
a Compare the rates at which carbon dioxide and methane are rising.
b Give reasons for the change in these greenhouse gas levels (see Table 4.1).

9a Using Figures 4.8–4.10, describe the general trend of changes to greenhouse gases in the atmosphere.
b With Table 4.1 to help, predict the implications of this trend.

by-products of the burning of fossil fuels. There are periodic changes in the amounts of these gases that affect the temperature of the atmosphere. It is this global warming that is really an enhanced greenhouse effect (see section 2.3).

Atmospheric pollution and global warming

Since the Industrial Revolution of the 19th century, human activities have resulted in rising concentrations of all the greenhouse gases. The main source has been the burning of fossil fuels and, more recently, the destruction of forests. To understand the possible human causes of global warming it is useful to look at the sources of the greenhouse gases in more detail (see Table 4.1). These gases are collectively important as they often absorb radiant energy in different parts of the energy spectrum.

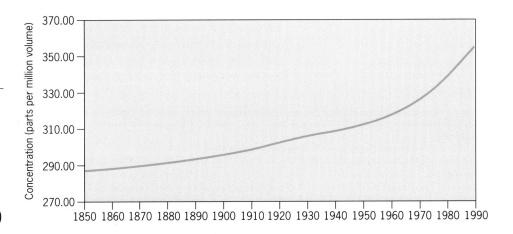

Figure 4.8 Changing concentrations of carbon dioxide, 1850–1950 (*Source:* CRU University of East Anglia)

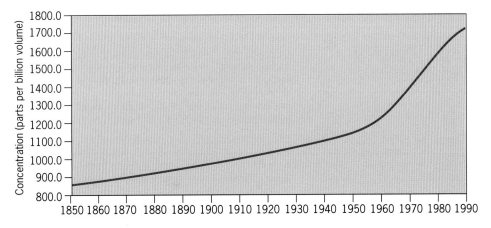

Figure 4.9 Changing concentrations of methane, 1850–1990 (*Source:* CRU University of East Anglia)

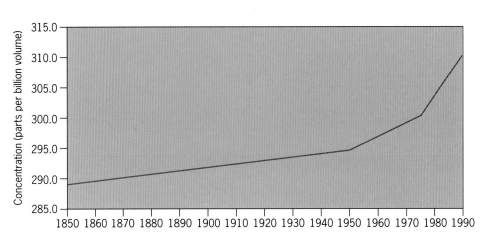

Figure 4.10 Changing concentrations of nitrous oxide, 1850–1990 (*Source:* CRU University of East Anglia)

?

10 Draw two flow diagrams showing examples of both negative and positive feedback in the carbon cycle (Fig. 4.11).

The carbon cycle

The carbon cycle is very complex and many areas where carbon moves from one part of the cycle to another are still not fully understood (Fig. 4.11). There are huge amounts of carbon in the atmosphere, in the oceans and on land that are exchanged naturally between these systems. For centuries this cycle was broadly in balance. *Outputs* from processes like respiration and the weathering of carbonate rocks were compensated for by processes like photosynthesis that absorb CO_2 in natural *sinks*. Similar amounts are exchanged from the oceans and seas each year.

The role of the oceans

Some scientists believe that oceanic CO_2 might be a key to understanding the fluctuating temperatures of the earth (Gribbin, 1988). A bloom of marine life, caused by a rise in temperatures or more sunlight, may lead to the removal of large quantities of atmospheric CO_2. This CO_2 falls to the sea floor when marine organisms die, causing an imbalance in the system. As a result, climate will cool due to the loss of the warming influence of the atmospheric CO_2.

In the long term, though, this may be offset by the fact that warm water absorbs less CO_2 than cold. More CO_2 will therefore be left in the atmosphere to encourage warming. What is not understood is whether the climate change triggers changes in the carbon cycle, or whether changes in the oceanic carbon cycle cause a change in climate. Other variables are also involved. For example, scientists suggest that, during an ice age, iron from wind-blown dust may reach the oceans. This nutrient could encourage the growth of plankton, leading to further removal of atmospheric CO_2.

Coral is also important. We have seen that the world's oceans are vital sinks or *stores* for CO_2. As the oceans warm, and their ability to absorb CO_2 decreases, more of the gas is left in the atmosphere. This in turn encourages global warming.

Figure 4.11 The carbon cycle

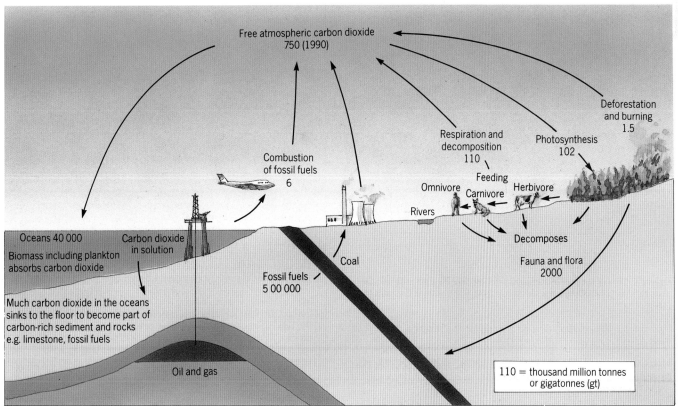

Carbon cycle

?

---

11 Describe and account for the trend on the graph in Figure 4.12.

12 Evaluate the two different views in Figures 4.13–4.14. How do you think governments should respond to contrasting advice of this nature? Give reasons for your suggestions.

However, as corals grow in warmer oceans, they take in CO_2. In fact, if sea levels change, some of the safest places might be islands like the Maldives and the Seychelles as coral growth should be able to keep pace with changes in sea level. In many other areas, though, like south-east Asia, human activity is destroying coral reefs. This has the double effect not only leading to a release of CO_2, but also reducing the role of coral reefs as a sink.

While scientists continue with experiments and modelling, they are now also using satellites to watch for blooms of plankton and to monitor other relevant environmental changes. These remain major areas of research.

The role of human activities

Human activities have disturbed the balance in the carbon cycle by adding huge quantities of CO_2 to the atmosphere. By the early 1980s an estimated 20 thousand million tonnes of CO_2 was added to the atmosphere annually from the burning of fossil fuels. Increasingly in the 20th century, deforestation has contributed to quantities of CO_2 being released. This occurs in two ways. Firstly, through the destruction of huge areas of forest and vegetation around the world – largely to create land for agriculture. There is then a consequent decline in photosynthesis, so CO_2 that would otherwise be absorbed remains in the atmosphere. Secondly, when vegetation is burnt the combustion releases additional CO_2. Scientists estimate that deforestation accounts for about 20 per cent of the CO_2 added to the atmosphere each year from human activities.

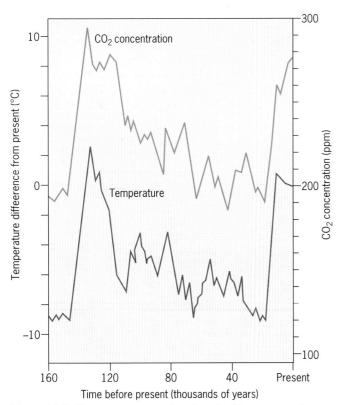

Figure 4.12 Air temperatures and carbon dioxide concentrations over the last 160 000 years, deduced from ice cores at Vostock in Antarctica

Figure 4.13 Why carbon dioxide is good for you (*Source: The Guardian*, 3 June 1993) (above right)

Figure 4.14 Is more or less carbon dioxide desirable? (*Source: The Guardian*, 5 June 1993) (below right)

After much number crunching, a computer in the United States endorses what some people have been muttering for years. If the greenhouse world predicted for the 21st century comes to pass, it may well be more fruitful than ours. Although environmentalists sometimes speak of carbon dioxide 'pollution', the life-giving gas is directly harmful only in suffocating concentrations. Should the tiny traces present in the air double, from roughly 300 molecules in a million to 600, the plants will love it. With biological productivity increasing by perhaps a quarter, the human species should benefit, at least on average.

This good news about carbon dioxide ought to suprise no one who knows that plants grow by reacting the gas with hydrogen, or that growers boost the carbon dioxide in their glasshouses to get higher yields. Plants also need water, and they like warmth… The climate forecasters say that doubling the carbon dioxide (or equivalent greenhouse gases) will increase global rainfall about 10 per cent, as well as warming the world by a few degrees Celsius.

Nigel Calder

Why does Nigel Calder assume that a possible increase in plant growth due to increased atmospheric concentrations of carbon dioxide is necessarily 'good news' for human beings? Such 'more is good' value judgements are premature. Since carbon dioxide released from the burning of fossil fuels is little able to distinguish between wheat and 'weeds', it is oversimplifying the preliminary results of US research to predict that heightened 'biological productivity' will be 'beneficial'. Nigel Calder fails to address the most important ecological change – the speed at which it happens. Human economic, social and agricultural systems may well be unable to adapt at the rates required by the changing climate and its terrestrial and oceanic consequences.

We could wait and see, hoping that Nigel Calder's scenario proves correct. Or we could take precautionary action now to cut carbon dioxide emissions (and many other environmental problems) by curbing traffic growth, improving energy efficiency and developing renewable energy systems. By so doing, we could reduce the extreme and known risks.

Simon Roberts, *Friends of the Earth*

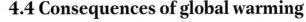
4.4 Consequences of global warming

The consequences of global warming are unclear and have caused much controversy. For example, we have all heard sensational claims which bear little relation to scientific research and often only add to confusion (see Figs 4.1–4.2). In fact, climate models are not yet accurate enough, and are unlikely ever to reliably predict climate change. Nevertheless, scientists generally agree that some climate change induced by human activities is certain.

The IPCC's 1990 report indicates that temperature rises will not be uniform throughout the world (Fig. 4.15). The IPCC also suggested that temperature changes will affect **precipitation** and soil moisture which in turn will seriously affect agriculture in many parts of the world. Agricultural output will be likely to fall in areas like the prairies of the USA and Canada and productive agriculture may shift to higher latitudes. Droughts may become more severe in the tropics. Global warming will also have important effects on many regions with major environmental, social, economic and political consequences (Fig. 4.16). It is important to remember, though, that the consequences of global warming are clouded by the uncertainty over the degree of climate change itself. A major problem concerns the lack of understanding of feedback mechanisms that could either encourage or dampen change in the atmospheric system. The feedback mechanisms themselves are likely to be interrelated, which complicates matters.

Cloud formation has an important influence on temperature (see sections 2.4 and 3.4). One possible negative feedback would involve increasing temperatures encouraging **evaporation**. Although **latent heat** (see section 2.4) would be released to the surrounding air when **condensation** occurred, cloud formation would more than compensate for this. Clouds would reflect an increasing proportion of **insolation** and cloud cover also has a shading and cooling effect.

Plankton in the oceans can also encourage cooling, not only through removal of CO_2 (see section 4.3) but also through the release of a chemical called dimethyl sulphide. When this chemical oxidises it creates **condensation nuclei** that further encourages condensation and cloud formation.

?

13   Study Figure 4.15.
a   Describe how temperatures differ across the world.
b   Which latitudes are most affected?

14a   Use Figures 4.15 and 4.16 to predict the impact of global warming on different parts of the world.
b   Suggest how people can prepare for these changes.

15   Construct two diagrams showing:
a   how clouds have a negative feedback influence on global temperature change.
b   how increasing levels of water vapour in the atmosphere could encourage global warming through positive feedback.

Figure 4.15 Global pattern of temperature rise at equilibrium with atmospheric concentration of carbon dioxide doubled (*Source:* Meteorological Office, GCM)

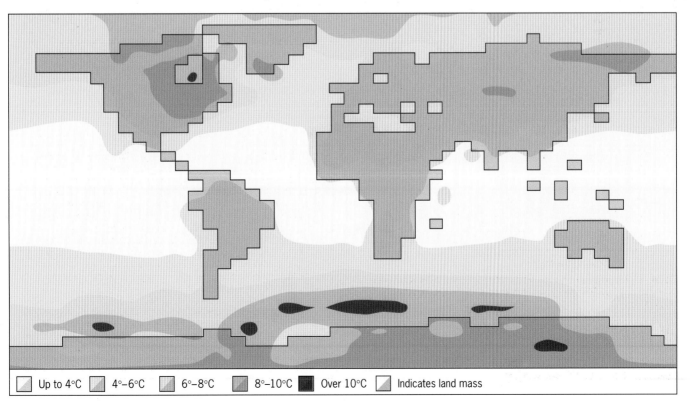

| | Up to 4°C | | 4°–6°C | | 6°–8°C | | 8°–10°C | | Over 10°C | | Indicates land mass |

?

16 Many of the world's major cities with coastal locations would experience severe disruption from a rise in sea levels. Use an atlas and Figure 4.16 to name such cities.

17 About one third of the Netherlands is below sea level. Carry out some research and study the likely consequences of a rise in sea level. See what you can find out about the protective measures that have been built along the Dutch coast to prevent flooding.

18a Use Figure 4.16 to assess the contrasting impact on richer and poorer nations if a 1 m rise in sea level (IPCC) occurs by the end of the next century.
b Suggest what steps need to be taken nationally and internationally against such a rise.

Figure 4.16 Predicted events if the earth's surface temperature increased by an average of 1°C (After: CollinsLongman *Environment Atlas*)

Increases in evaporation could add to the warming effect as water vapour is a greenhouse gas; an example of a positive disturbance to the system. These relationships between evaporation, temperatures and cloud formation are still not fully understood.

Sea levels

One effect of global warming will be a rise in sea levels – approximately half brought about by melting glaciers and ice sheets and half by the thermal expansion of water. In fact, there are indications that this has already started, with an estimated sea level rise of about 15 cm in the 20th century. The IPCC predicts a further rise of 18 cm by 2030 and 44 cm by 2070.

Low-latitude glaciers have added only a relatively small amount of this water to date. In contrast, the future of the polar ice caps is more critical. The British Antarctic Survey (BAS) recently reported the fastest sustained temperature rise since worldwide temperature records began 130 years ago (Fig. 4.17). BAS also reported the break-up of the Wordie Ice Shelf with the loss of about 800 km² of pack ice between 1969 and 1989 (Fig 4.18). There have been exaggerated claims that the west Antarctic ice sheet may break up and slide into the sea causing a massive rise in sea levels and consequent flooding. Scientists at BAS argue, though, that this is unlikely for about a century. If this does occur, it will span centuries rather than decades and raise sea levels by 5 m at a rate of about 2 cm a year.

These fluctuating temperatures at the Faraday Base may, however, only be temporary variations. It is the small changes in the amount of pack ice at the edges of the ice shelves which causes large temperature changes. It is also likely that global warming could encourage growth of the polar ice caps. This is because increases in evaporation at low latitudes could lead to increasing condensation and

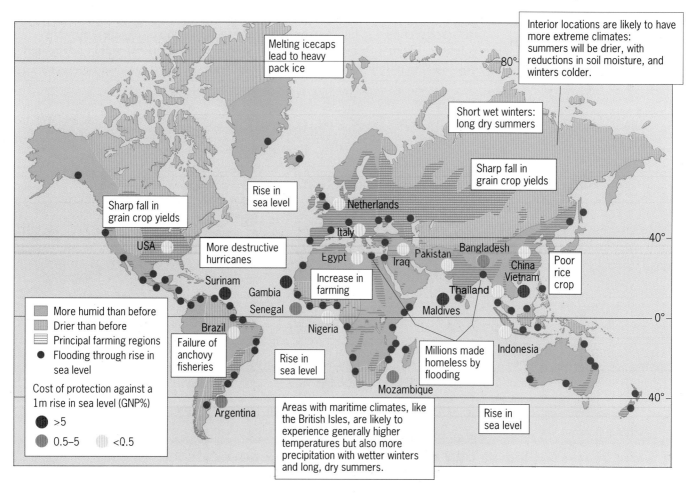

19 Study Figure 4.17. Estimate the average difference in temperatures between 1945–50 and 1985–90.

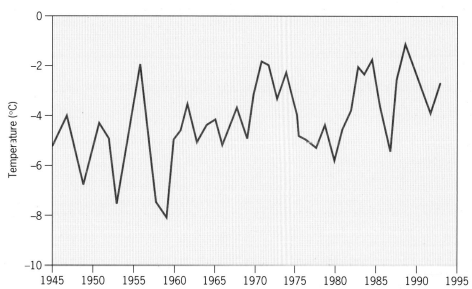

Figure 4.17 Mean annual temperature, Farraday Base, Wordie Ice Shelf, Antarctica

Figure 4.18 Landsat image of the Wordie Ice Shelf, Antarctica. Between 1974 and 1989 there has been a dramatic retreat of the Wordie Ice Shelf. Fracture processes have weakened the shelf and led to a retreat of the ice front.

cloud cover, greater reflection of insolation, and so more precipitation at the poles with an accumulation of snow and ice.

Clearly, low-lying areas of the world are particularly vulnerable to rising sea levels, such as the Netherlands and Belgium and the delta environments of many of the world's major rivers e.g. the Nile, Mississippi and the Ganges. A rise of about 1 m could lead to the virtual disappearance of many low islands, such as those of the south Pacific and the Maldives in the Indian Ocean. This in turn would lead to population displacement, damage to communications and the salinisation and flooding of agricultural land. There would also be consequences for recreation and tourism as destinations like the Seychelles shrink and coastal environments change.

Population and food

A study commissioned by the US Environmental Research Agency indicated that in the 21st century global warming could add hundreds of millions of people to those who already suffer from hunger. The UN estimates about one tenth of the world's population already has an inadequate diet and is undernourished. Even without climate change, population growth could double this by 2100.

Three computer simulations predicted how the world's climate would change by 2060 based on current emissions of greenhouse gases. They all showed increases in temperature and changes in rainfall. One simulation was by the UK Meteorological Office. Estimates were made of how these changes would affect yields of staple crops. Another program simulating world food trade was used to predict the effect on food prices, food distribution and numbers of hungry people.

Although the study showed that global warming might improve crop yields in the wealthy, high-latitude countries like Canada, this was more than offset by declining yields in the poorer tropical and sub-tropical countries that also have the bulk of the world's population. One of the main findings was that world cereal production was predicted to be 1–7 per cent lower in 2060 than it would be without human-made climate change.

These predictions show the need not only for global agricultural strategies but also for more attention to be given to population growth.

Other consequences

There will undoubtedly be many other results of global warming with various impacts on the way we live (Table 4.2).

Table 4.2 Selected consequences of global warming on our economic and social well-being

Environment	Example of impact	Consequences
Landscape	Glaciers may disappear in Alps. Reduction in snow cover.	Tourism affected by reduction in winter sports.
Hydrological	In areas of reduced or irregular rainfall, wetlands will shrink, rivers be reduced in size and waterfalls dry up. Reduction in rainfall will influence water supplies e.g. underground aquifers.	Water authorities will have to develop engineering plans to cope with the imbalance between demand and supply, even if this is only a seasonal problem. Irrigation, hydro-electric power schemes and navigation will be affected.
Maritime	Changes to the temperature of the sea and the position of ocean currents will cause changes in the location of wildlife.	Fish numbers and the location of shoals could change with implications for the fishing industry.
Meteorlogical	Changes in the position of the **jet stream** (see section 6.5): A shift south as a result of higher temperatures would cause depressions to move into warmer waters and so intensify into powerful storms. **Hurricanes** are also likely to be stronger if global warming continues.	More intense storms create problems for weather forecasters who would have less time to predict the storm's course over Europe e.g. the severe damage caused by the storm which hit the British Isles in October 1987 (see section 6.5). It is significant that three of the most severe hurricanes on record occurred in the Caribbean since 1988 e.g. Hurricane Gilbert (1988) and Hurricane Andrew (1992) were both Grade 5 storms.
Regional weather	Interior locations, e.g. Great Plains of North America, to have more extreme climates. Summers will be drier with reductions in soil moisture; winters colder. Areas with maritime climates, e.g. British Isles, to experience higher temperatures with wetter winters and long, dry summers.	The uncertainties about the degree of global warming that might occur mean that it may be several decades before we can be more certain of how regional weather and climate will change.
Vegetation	Distribution of flora and fauna will shift to different latitudes. Species vulnerable to change may become extinct e.g. great raft spider in Britain.	Many areas may benefit from new capacity to grow different crops e.g. aubergines or melons in Britain. Pests will spread. Changing crop yields may affect a growing population.
Human health	The range of tropical diseases, e.g. malaria and cholera, will increase as temperatures rise.	Death rate may increase in some parts of the world. Increased pressure on medical services.

20 Using Figures 4.15, 4.16 and Table 4.3, suggest how those countries listed in Table 4.3 might:
a be affected by global warming.
b be able to cope with the changes brought about by global warming. Use Table 4.2 to help.

Table 4.3 Socio-economic measures for countries at different stages of development

Country	Per capita GNP US$ 1991	Per capita energy consumption million tonnes coal equivalent 1990	Economically active adults employed in agriculture 1991 (%)	Adult literacy 1991 (%)	Food supply: calorie intake per person per day 1988–90 (average)	Population per doctor 1991
Australia	16590	7.53	5	99	3302	400
Canada	21260	10.26	3	99	3242	450
Egypt	620	0.74	40	48	3310	1320
Jamaica	1380	0.83	27	98	2558	2040

The impact of global warming on the British Isles

Many theorists believe that global warming would result in the British Isles having a more Mediterranean climate. If this did occur, not only would our climate change, but it would radically alter many aspects of our lives (Fig. 4.19).

Changes to crop belts

A warmer climate would result in substantial changes to agriculture (see Fig. 4.19). Soil erosion might increase in a warmer climate. The drizzle and frequent rain we currently experience would be replaced by less frequent heavy storms, like those that occur in southern Italy.

Changes to wildlife and vegetation

Changes in vegetation and climate would also affect wildlife. Nine organisations, including the Ministry of Agriculture, Fisheries and Food (MAFF) and the Forestry Commission, are sponsoring an Environmental Change Network. This consists of eight research centres designed to monitor long-term changes in flora and fauna. Entomologists believe that global warming could lead to outbreaks of many insect pests while other species would disappear from the English countryside. More seriously, a warmer climate would encourage the spread of malaria as

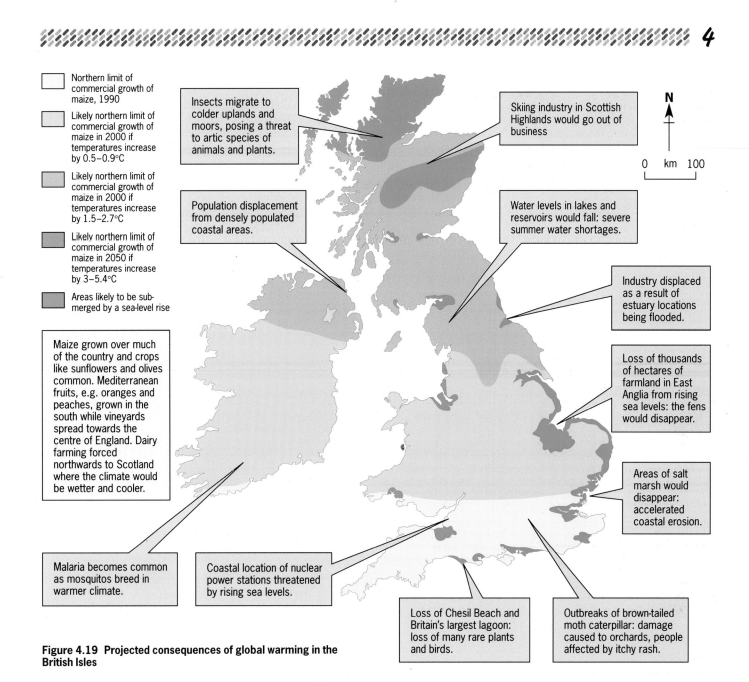

Northern limit of commercial growth of maize, 1990

Likely northern limit of commercial growth of maize in 2000 if temperatures increase by 0.5–0.9°C

Likely northern limit of commercial growth of maize in 2000 if temperatures increase by 1.5–2.7°C

Likely northern limit of commercial growth of maize in 2050 if temperatures increase by 3–5.4°C

Areas likely to be sub-merged by a sea-level rise

Insects migrate to colder uplands and moors, posing a threat to artic species of animals and plants.

Skiing industry in Scottish Highlands would go out of business

Population displacement from densely populated coastal areas.

Water levels in lakes and reservoirs would fall: severe summer water shortages.

Industry displaced as a result of estuary locations being flooded.

Loss of thousands of hectares of farmland in East Anglia from rising sea levels: the fens would disappear.

Maize grown over much of the country and crops like sunflowers and olives common. Mediterranean fruits, e.g. oranges and peaches, grown in the south while vineyards spread towards the centre of England. Dairy farming forced northwards to Scotland where the climate would be wetter and cooler.

Areas of salt marsh would disappear: accelerated coastal erosion.

Malaria becomes common as mosquitos breed in warmer climate.

Coastal location of nuclear power stations threatened by rising sea levels.

Loss of Chesil Beach and Britain's largest lagoon: loss of many rare plants and birds.

Outbreaks of brown-tailed moth caterpillar: damage caused to orchards, people affected by itchy rash.

Figure 4.19 Projected consequences of global warming in the British Isles

those mosquitoes thrive in Mediterranean conditions.

Rising sea levels

The Climatic Research Unit at the University of East Anglia has estimated that sea levels could rise by as much as 38 cm by 2030. A more likely figure, though, is 15 cm. However, Britain is still responding to loss of ice at the end of the ice age. As a result, the north is rising slightly, while the southern part is sinking. This would only serve to worsen the effects of a rise in sea level. The Institute of Terrestrial Ecology has estimated that £5 billion would have to be spent on building the additional defences to prevent the flooding that could result from global warming. By 2030, it might be necessary to rebuild the Thames Barrier completed in 1984 to protect London from flooding (see Fig. 9.18).

21 If the south of England were to become more like southern Italy and the north like the central-west of France, use Figure 4.19 and an atlas to describe what the climate of these two parts of the country would be like.

22 What implications could changes in vegetation have for people who suffer from allergies and hay fever?

23 Classify under suitable headings the likely impacts of a rise in sea level on the coastal areas of the British Isles (Fig. 4.19).

24 Make a list of ten different physical features/environments in Britain and describe and explain how they could be affected by global warming. Examples might include the wetlands of the Somerset Levels, Spurn Head near Hull or Chesil Beach on the south coast.

25 Suggest what responses we need to take (if any) if global temperature rises are solely a natural process.

26 Consider the view of the Earth Summit. Do you agree with this statement? Give your reasons.

4.5 What can be done?

To answer this question we need to return to the two hypotheses raised at the beginning of section 4.2. If we have not been able to prove that global warming is taking place, then we do nothing. However, doubt and conflicting views at the Earth Summit in 1992 gave birth to the *precautionary principle*. This states that 'where there are threats of serious damage, lack of full scientific certainty shall not be used as a reason for postponing cost-effective measures to prevent environmental degradation'. There is very strong evidence that human activities are contributing to global warming, in which case we need to find out what precautionary steps we can take.

Reduction of greenhouse gases
Attempts to reduce emissions of greenhouse gases have largely concentrated on CO_2. However, we have seen earlier that it is also necessary to target other gases (Table 4.4).

Table 4.4 Measures to reduce greenhouse gases

Measure	Example	Pros and cons
Changing use of fossil fuels	Change from the heavy use of coal to oil and gas. Gas: • Modern gas turbine power stations produce for each unit of electricity generated less than 50% as much CO_2 as a modern coal-fired power station. • Burning gas is also more energy efficient (approx. 60% is turned to energy instead of about 45% from a coal-fired station).	• The British government's commitment to stabilising CO_2 emissions at 1990 levels by the year 2005 partly depends on substituting gas for coal. • This could involve the construction of 14 new power stations.
Alternative energy sources	• Renewable energy sources, e.g. HEP, tidal, wave, solar, and wind power, would reduce CO_2 output. • Burning methane (a potent greenhouse gas) at landfill waste sites. • An increase in nuclear power.	CO_2 is also released when methane is burnt. Nuclear power (view of British Nuclear Fuels): • As a result of producing 66% of their electricity by nuclear power, France and Belgium have reduced their CO_2 output faster than any other country in Western Europe. • Investment is encouraged into nuclear power (BNFL invested £1.5m a day at Sellafield in 1989). Nuclear power (view of Friends of the Earth): • To satisfy demand, one nuclear power station needs to be built every 12 days over the next 37 years, at great expense. • An increase in nuclear waste with no effective means of disposal.
Energy pricing and efficiency	• Increasing the fuel efficiency of buildings. • Increasing the fuel efficiency of transport. • Global energy pricing strategy that involves pricing fuels to reflect their environmental costs.	Buildings: • In the UK, the DoE introduced new standards in building regulations in 1990: new buildings are to have a 20% improvement in energy efficiency. Transport: • Persuading governments to adopt policies that encourage the use of public transport and the use of railways (rather than roads) for freight transport. • Many car and aircraft companies are developing more fuel-efficient engines (to increase sales). • Fossil fuels would be more expensive and the polluter would pay more as a penalty for greenhouse gas emissions.
Control of deforestation	• Provision of constructive advice on forest management.	• Difficult to persuade e.g. the Brazilians to control the destruction of the rainforests of Amazonia. • The boreal forests of the higher latitudes also need to be managed effectively.
Programmes to replant trees (afforestation)	• Alternative uses like tourism could help to conserve these resources.	All forests not only act as a sink for CO_2 but can also help with other consequences of global warming, like soil erosion and the management of watersheds.

27 Consider Table 4.4.
a In groups, discuss the pros and cons of nuclear power as a way of influencing levels of greenhouse gases.
b Analyse the value positions people in your group hold.
c In Table 4.4 two organisations' views are presented. Which do you think:
• does, and
• should have, the greater influence on government policy? Give reasons.
d State your own value position and outline the energy policy you think the government ought to adopt to reduce emissions of CO_2. Give reasons.

28 It seems that we may all have responsibilities towards global warming. Outline what steps an individual can take to reduce emissions of greenhouse gases.

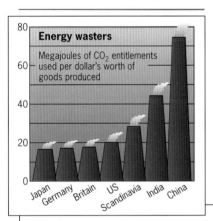

Adapting to a changing environment

The changes brought about by global warming will probably be gradual, thus giving us time to adapt. For example, it is predicted that the flooding of low-lying coastal areas will take place over many decades. This will allow a gradual displacement of population and changes to settlements and infrastructure. However, the IPCC (1992) argues that in a densely populated country like Bangladesh, it is the rapid population growth that would create pressures on available land and resources more swiftly than any flooding associated with global warming.

International efforts

The problems brought about by global warming, the changing patterns of food production and lack of food in some areas could be severe. If droughts and famines worsen in low latitudes, there could be huge population movements with consequences for other parts of the world.

Clearly, there will be a need for international co-operation to cope with such major global problems (Fig. 4.20). We have seen that the UK has been taking steps to apply technology and economic measures to combat global warming. Although the efforts of individual countries are vital, if the output of greenhouse gases is merely transferred elsewhere there is no net global benefit. This particularly concerns the economically developing countries because, as they seek to industrialise and develop their economies, they will inevitably consume more energy. However, the economically developed countries cannot impose low levels of development on the developing world on the grounds of global warming. Transfers of technology will be required from richer to poorer countries and trade practices introduced to help their development. This will have consequences for companies in the developed world that may have to accept competition from newly industrialising countries. As consumers, we may have to accept higher prices and changes to our standard of living which reflect true environmental and social costs.

Agreements

The IPCC has already encouraged research and published the results of scientific studies. One outcome from the IPCC predictions on global warming (see section 4.2) was the signing of a treaty at the United Nations Conference on Environment and Development (UNCED) at Rio de Janeiro in 1992. The World Convention on Climate commits the 160 nation signatories to energy-saving programmes aimed

Greenhouse warming goes to market

BROKERS in the City and on Wall Street could soon be dealing in permits to emit greenhouse gases, if UNCTAD, the UN Conference on Trade and Development, gets its way. Permission to add to global warming could be bought and sold like coffee futures or government bonds.

Last week at a meeting in Prague, UNCTAD called on the US, the European Union and Japan to set up a pilot pollution 'exchange', to organise trade in 'emissions entitlements' for carbon dioxide. These countries contribute 40 per cent of the world's carbon dioxide emissions from burning fossil fuels. Under the Climate Change Convention, which holds its first formal meeting in Berlin in March, they are all required to reduce their carbon dioxide

emissions to 1990 levels by the year 2000.

A market would allow governments to allocate part of their pollution entitlement to individual polluting companies, such as power generators. The companies could then buy and sell the permits. Dirty companies that emitted too much carbon dioxide would have to buy permits from clean ones.

At last week's meeting, Frank Joshua, head of UNCTAD's global interdependence division, also called on poorer countries to join in as soon as possible and pave the way for a global exchange system. This could happen after 2000, when the Climate Change Convention may set emissions targets for some developing countries. If poor nations received more permits than they required, they would be able to sell

unused permits to industrialised nations that were unable to meet their targets.

Economists say the system would provide an incentive for investment in cleaner technologies. A country that developed wind power or a better railway network, for example, might be able to recoup some of the costs by selling excess pollution permits to countries still stuck with coal-fired power stations and heavy road traffic.

Richard Sandor, an economist and director of the Chicago Board of Trade, who is one of the authors of the proposals, believes that 'air and water are simply no longer free goods. They must be redefined as property rights so that they can be efficiently allocated'.

Figure 4.20 Carbon trading system (*Source:* Fred Pearce, *New Scientist*, 21 January 1995)

Last try for Europe-wide climate agreement

Environment ministers of the European Union meet in Brussels this week for a final attempt to reach agreement on jointly ratifying the Climate Convention they signed in Rio in 1992. Under the convention, they pledge to stabilise their emissions of carbon dioxide at 1990 levels by the year 2000.

But the promises they made in Rio are looking increasingly empty. Britain has already scuppered plans to share the burden of reducing CO_2 emissions among members of the EU. And Jacques Delors, president of the European Commission, has proposed a programme of road building that will inevitably lead to more traffic – Europe's fastest-growing source of CO_2 – and to increased energy consumption.

Figure 4.21 Attempts to ratify the Climate Convention (*Source:* Deborah MacKenzie, *New Scientist*, 18 Dec, 1993)

29a How does Figure 4.21 suggest that some nations were not entirely serious about their intent to control greenhouse gases?
b How does this reflect different values towards cuts in emissions of greenhouse gases?

30 Read Figure 4.20. Assess the strength of the emissions exchange system. How successful do you think it could be? Justify your response.

31 Essay: Explain what you understand by the term global warming and outline the main causes.

32 Essay: What are likely to be the main consequences of global warming at the world scale? Consider the options available to reduce the impact of global warming and assess the costs of these to individuals, corporations and countries. Make your own recommendations for tackling global warming and give reasons for your suggestions.

at reducing global warming. However, there was no commitment to particular dates or targets, so the treaty was only a statement of intent rather than a plan for action (Fig. 4.21). Many of the richer countries objected to setting aside money to help poorer countries and refused to cut back their own emissions. More than 50 countries have now formally approved their signatures, and the first conference of the Parties to the Convention took place early in 1995 to discuss strategies to reduce emissions.

The Montreal Protocol also resulted in a reduction of CFC emissions and a further agreement in 1990 resulted in a complete ban by the year 2000. Although these initiatives arose out of concern for damage to the ozone layer, CFCs are also greenhouse gases (see Table 4.1).

Whatever doubts there might be about global warming, the pace of change, the causes or the consequences, one of the greatest dangers is overreaction. We have considered alarmist, unscientific responses, but if global warming exists, it is equally dangerous to bury our heads in the sand and hope the problem will go away. Research will need to continue monitoring the environmental variables involved. A supplement to the IPCC's 1990 report (referred to previously), published in 1992, indicated that our understanding of global climate was still inadequate to produce accurate global climate models. The report stated the following areas were still poorly understood:

1 The link between global ice sheets and global warming.
2 The heat absorbing and heat retention properties of the oceans and how this influences ocean circulation.
3 The feedback effect of clouds on greenhouse gas-induced global warming.
4 The effects of feedback from hydrological and ecological processes on land.

It may be several decades before we can be more certain about the way the world's climate is changing. It is important, though, not to let global warming detract our attention from other severe environmental problems. In the meantime, one option is to adopt the precautionary principle and take appropriate steps to combat global warming. This would include energy conservation and controlling fossil fuel consumption resulting in many other environmental benefits – although at a high economic cost. Another option is to avoid extra costs and run the risk that global warming might not happen. We would then have to accept the consequences.

Summary

- An apparent sudden rise in global temperatures, believed to have been caused by human activities, has increased our concern over global warming.
- There has been a great deal of media hype over global warming and many unscientific arguments about the causes and consequences. It is therefore necessary to adopt a questioning approach and to examine the evidence carefully.
- There are a number of greenhouse gases. The principal ones, in terms of their present contribution to global warming, are water vapour and carbon dioxide. Changes in the levels of these gases may be contributing to global warming. The earth is getting warmer and the levels of these gases have increased, but as yet there is no direct link.
- There are a number of natural sources of greenhouse gases, as well as the contributions made by human activities.
- Although climate models are not yet able to accurately predict climate change, scientists generally agree that a level of climate change is certain. There are indications that climate change will not be even over the globe.
- Broadly, there are three active responses to global warming. We can take steps to reduce emissions of greenhouse gases. We will also have to adapt to a different environment. Thirdly, the scale of the problem is such that the steps that need to be taken, including research efforts, will require international co-operation.

5 *Ozone depletion*

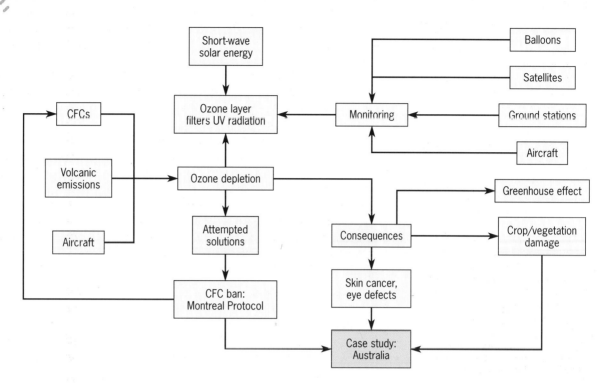

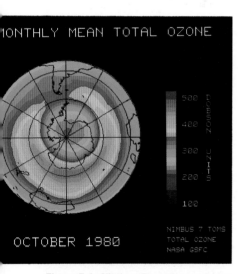

Figure 5.1 Nimbus-7 weather map of the hole in the ozone layer over Antarctica, October 1980: the hole largely avoids populated areas. The colours represent total ozone concentrations (measured in dobson units) running from purple for the lowest concentration, through blue, green, yellow and orange for the highest.

5.1 Introduction

About 20 km above us in the earth's **stratosphere** is an area of ozone gas at greater concentrations than anywhere else in the atmosphere (see section 2.3). The **ozone layer** filters out inputs of ultraviolet short-wave radiation from the sun. In the early 1970s scientists recognised a thinning of the ozone layer. This thinning accelerated in the 1980s resulting in a hole over Antarctica about the size of the USA, nine million km^2 in area (Fig. 5.1). This 'hole in the sky' (as it is called in New Zealand) causes us concern because increased ultraviolet (UV) radiation can lead to skin cancer and eye defects like cataracts.

Although initially there was some disagreement over how serious the ozone depletion problem was, this is no longer the case (Fig. 5.2). Many people believe the cause of ozone depletion is largely the result of human disturbance to the atmospheric system, the greatest issue being the use of *chlorofluorocarbons* (*CFCs*). We will try to establish the extent to which ozone depletion is a natural or human phenomenon. Whichever it is, scientists now recognise that continued depletion of the ozone layer could pose a serious threat to life in many parts of the world.

5.2 The nature of ozone and the ozone layer

Our planet is probably unique in the solar system for having an atmosphere that is chemically active and contains oxygen. Life would not exist without oxygen and, equally, oxygen is dependent on living plants on earth. During a process covering billions of years, oxygen became available in the earth's atmosphere (Fig. 5.3). Eventually, oxygen was converted into ozone, a gas that absorbs UV radiation. The ozone layer gradually developed and today forms a layer of the gas in the stratosphere (see section 2.3). This layer of ozone in the atmosphere acts as a protection, without which it is doubtful if there would be any life on earth (Fig. 5.4).

?

1a Before studying ozone depletion in this chapter, write down your own understanding of this issue.

b Study Figure 5.2 and list the concerns over depletion of the ozone layer.

c Compare your own views with those in Figure 5.2 and suggest reasons for any differences.

2 Using Figure 5.1 and an atlas, identify the more densely populated areas that will be particularly under threat if the hole over Antarctica increases in size.

It is now far beyond doubt that chlorofluorocarbons (CFCs), halons and related chemicals are causing massive depletion of the ozone layer.

The weakened ozone layer could result in millions of extra cases of skin cancer, of cataracts and of infectious diseases. The onset of AIDs may become more rapid after infection with HIV. Enormous damage may also be done to the natural environment, ranging from losses in crop yield to the potential collapse of the marine food chain and the alteration of the planet's climate.

Given the risks to human health and the environment, further damage to the ozone layer must be prevented. The use of ozone-depleting chemicals must be phased out without delay.

Figure 5.2 Pressure group information about ozone depletion (*Source*: Friends of the Earth, 1992)

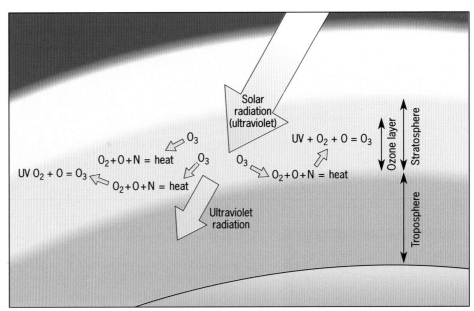

Figure 5.4 How the ozone layer intercepts ultraviolet radiation

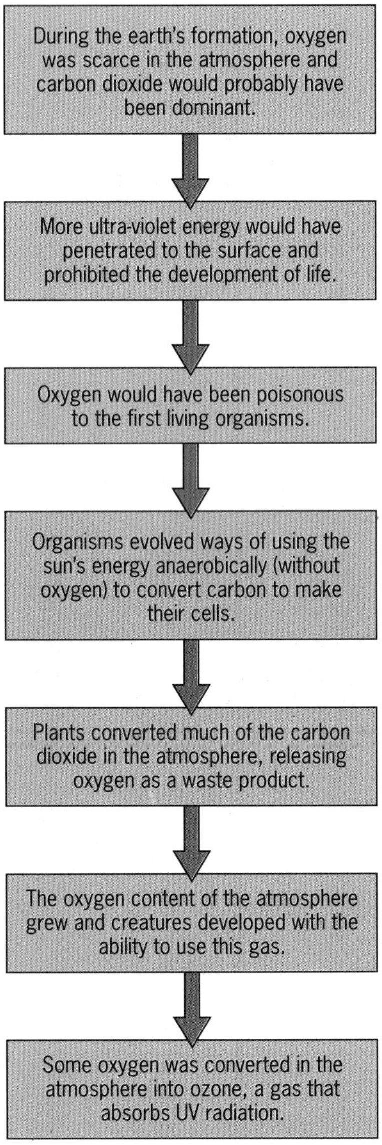

During the earth's formation, oxygen was scarce in the atmosphere and carbon dioxide would probably have been dominant.

More ultra-violet energy would have penetrated to the surface and prohibited the development of life.

Oxygen would have been poisonous to the first living organisms.

Organisms evolved ways of using the sun's energy anaerobically (without oxygen) to convert carbon to make their cells.

Plants converted much of the carbon dioxide in the atmosphere, releasing oxygen as a waste product.

The oxygen content of the atmosphere grew and creatures developed with the ability to use this gas.

Some oxygen was converted in the atmosphere into ozone, a gas that absorbs UV radiation.

Figure 5.3 The history of oxygen and ozone

Ozone formation

The oxygen we breathe has two atoms in each molecule of the gas (O_2). Ozone is a form of oxygen with three atoms (O_3). In the atmosphere, some UV radiation provides the energy to break ozone into O_2 and O. This reaction can be reversed so that the single oxygen atoms combine with oxygen molecules when they absorb UV radiation – again to form ozone. The process of ozone formation will only happen, though, when the kinetic energy involved in the reaction can be taken up by another molecule: this is usually nitrogen (N). The N molecule moves faster with this additional energy, and as gas molecules move faster they become hotter. This is the process, then, by which ozone is formed and ultraviolet energy is absorbed to warm the stratosphere. Ozone is therefore constantly being destroyed and created in the atmosphere in a dynamic balance. This is contrary to the view that ozone is a resource, like coal, that is being used up.

The atmosphere contains small amounts of many gases other than oxygen. Chemical reactions occur naturally in the ozone layer that accelerate the breakdown of ozone (Fig. 5.5). For example, oxides of nitrogen convert O_3 back to molecules of O_2. Although ozone is constantly being made in the ozone layer, human activities can interfere with the process – leading to a reduction in ozone. Simply, the production rate of ozone remains the same but the destruction rate increases.

Figure 5.5 Ozone sources and interactions with other chemicals

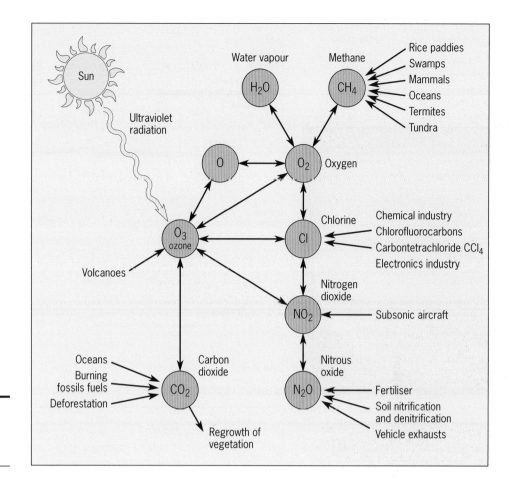

?

3 Use Figure 5.5 to explain how ozone is naturally both created and destroyed in the ozone layer.

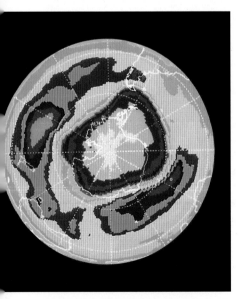

Figure 5.6 Meteor-O3 weather map of the hole in the ozone layer over Antarctica, October 1994: the hole affects parts of Chile, Argentina and Australia. The colours represent total ozone concentrations (measured in Dobson units) running from grey for the lowest concentration, through pink, yellow, green and brown for the highest.

5.3 Ozone depletion and its causes

Antarctica

Research scientists from the British Antarctic Survey (BAS) first publicised their research of the hole in the ozone layer over Antarctica in 1985 (see Fig. 5.1). Joe Farman and Jonathan Shanklin confirmed from their research that between 1955 and 1985 the amount of ozone over Antarctica had fallen by about 50 per cent. Scientists initially thought that ozone depletion would be very slow and that this would only be a problem in the 21st century. They supposed that reductions were in the order of only a few per cent over several decades. However, the discovery of the hole over Antarctica was disturbing; it seems that a threshold was reached beyond which ozone depletion accelerated. In the early 1990s satellites revealed that ozone loss was becoming more severe over Antarctica (Fig. 5.6). In addition, the CFCs which we use today are very inert, which means that they may last for 100 years in the stratosphere, continuing to break down ozone.

The hole in the ozone layer was for a long time largely restricted to Antarctica. It is at its largest each spring but a degree of 'repair' takes place over the next few months. The reason for this is the unique **meteorology** of this part of the world. In the Antarctic winter, the polar stratosphere is separated from the tropical stratosphere by a belt of strong westerly winds: the **polar vortex**. This largely contains the processes that lead to the development of the ozone hole (Fig. 5.7).

The Arctic

In the late 1980s scientists discovered disturbing indications of a hole developing over the Arctic. In 1989 and 1990 measurements in Switzerland showed that a thinning of the ozone layer was affecting Europe. Research at the University of

Figure 5.7 Stages in the development of the ozone hole over Antarctica

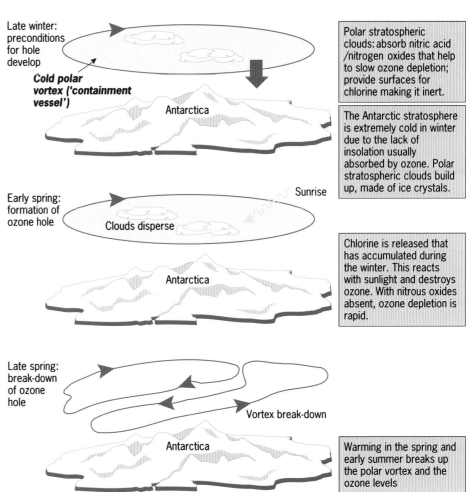

Polar stratospheric clouds: absorb nitric acid /nitrogen oxides that help to slow ozone depletion; provide surfaces for chlorine making it inert.

The Antarctic stratosphere is extremely cold in winter due to the lack of insolation usually absorbed by ozone. Polar stratospheric clouds build up, made of ice crystals.

Chlorine is released that has accumulated during the winter. This reacts with sunlight and destroys ozone. With nitrous oxides absent, ozone depletion is rapid.

Warming in the spring and early summer breaks up the polar vortex and the ozone levels

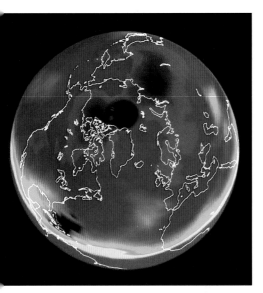

Figure 5.8 Nimbus-7 and Meteor 3-05 weather map showing the change in total ozone for the northern hemisphere between March 1993 and the March average for 1979–90. The colours represent the percentage difference: most of the northern hemisphere shows a loss in total ozone of 5–15% (blue), severe depletion is shown as purple (minus 15–25%), increase in red (equatorial belt). Black over North Pole indicates missing data due to polar night.

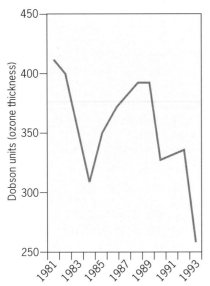

Figure 5.9 Average ozone thickness over the British Isles (measurements taken 5, 6, 7 March each year), 1981–93 (*Source:* Met. Office)

Innsbruck, in Austria, revealed an increase of one per cent in ultraviolet radiation over the Alps between 1980 and 1990. Some scientists argued, though, that such small changes could be due to chemicals that are associated with pollution from urban areas. These include nitrous oxide, methyl chloride, methane, fluorine and carbon tetrachloride which react under the influence of UV light to form compounds that speed up the breakdown of ozone. However, satellite images confirmed a developing hole over the Arctic in the early 1990s (Fig. 5.8). This is more serious than the hole over the Antarctic because the northern hemisphere is so much more densely populated. In 1991 there were reports that an eight per cent reduction in ozone over the last decade was affecting an area from northern Britain to southern Spain. In 1992 scientists announced a drop in ozone over Europe in the first two months of the year of up to 20 per cent. Similar falls were quoted over Canada (16 per cent in January) and Russia (15 per cent in February and March) while by 1993 measurements over the UK showed a disturbing degree of ozone depletion (Fig. 5.9).

Chlorofluorocarbons

Chlorofluorocarbons (CFCs) were first suggested as the cause of ozone layer depletion by two US scientists, Mario Mowlina and Sherwood Rowland, in 1974. They had studied a growing 'hole' in the ozone layer that had also been detected by American satellites. Their ideas were initially treated sceptically and rejected in many cases as being unbelievable. However, research since then, particularly by NASA and BAS, has led to wide acceptance of ozone depletion. The major cause of this depletion has been human-made, largely due to the use of CFCs which show a dramatic rise in production since the 1950s (Fig. 5.11).

Figure 5.10 Annual percentage change in total ozone coverage

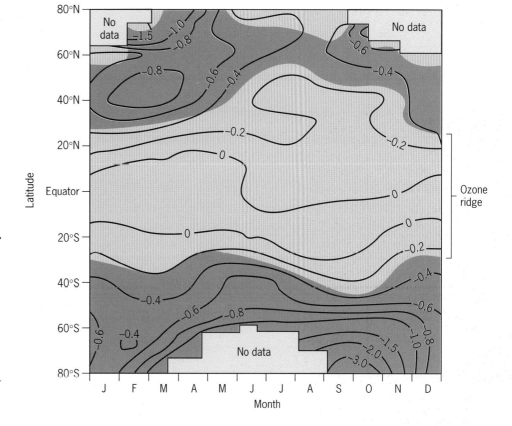

?

4a Use Figure 5.10 to estimate the percentage decrease in ozone over the lower latitudes including much of Africa, South America and Indonesia.
b Which areas of the world have not been seriously affected by ozone depletion?

Figure 5.11 Concentrations of CFC11 and CFC12, 1950–90

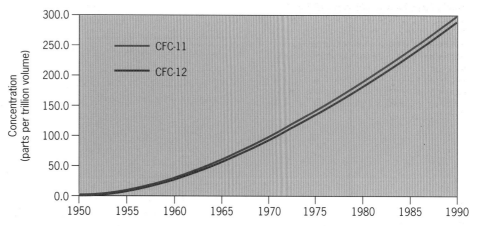

The main pollutant that reacts with ozone is chlorine. Chemicals called chlorofluorocarbons drift into the stratosphere and release chlorine which, through complex reactions, breaks down ozone (Fig. 5.12 and see Fig. 5.5). There are a number of different CFCs as well as other chemicals which destroy ozone (Table 5.1).

Natural causes

With all the attention focused on CFCs and similar chemicals, it is easy to overlook the natural causes of ozone depletion. Volcanic eruptions can also damage ozone, as volcanic dust has the effect of removing nitrogen oxides and increasing carbon monoxide in the atmosphere (see Fig. 5.5). These lower the ozone layer's immune system. Since about 1980 there have been a number of major volcanic eruptions, including Mt St Helens in 1980 and El Chicon in Mexico in 1982. Mt Pinatubo, in 1992, increased the concentrations of dust and carbon monoxide ten-fold in the stratosphere in middle and high latitudes (see Fig. 3.19).

Table 5.1 Chemicals damaging the ozone layer

Chemical	Use	Time of use
Chlorofluorocarbons		
CFC11	Foaming agent in making rigid or flexible plastic foam. Rigid foam is used to make fast food packaging which is light, strong and a good heat insulator e.g. trays for meat	1970s–1980s
CFC11, CFC12	Propellant in aerosol sprays	1960s–1980s
Other CFCs	To make furniture, wall insulation and as the working fluid in fridges	1960s–1990s
Other chemicals		
Halons	Fire extinguishers	Rarely used, but stored as a 'bank' in equipment
Carbon tetrachloride	In pesticides, pharmaceuticals and synthetic rubber	Released during manufacture of products
Trichloroethane	As a cleaning solvent for metal parts	Released during the manufacture of metal equipment

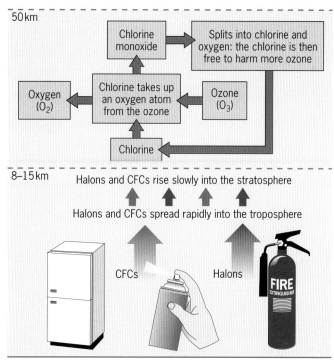

Figure 5.12 How CFCs and halons attack the ozone layer

Some fungi that rot wood are known to generate a chemical called chloromethane. This chemical is a source of about 25 per cent of stratospheric chlorine. Scientists have suggested that deforestation, resulting in less rotting wood and litter, causes reduced quantities of fungi. In turn, this would lead to a reduction in chloromethane and so the rising concentration of CFCs may not be so serious. More extensive research, however, is necessary to establish how this fits into the total picture of the chemistry of the ozone layer. Despite these natural contributions, the evidence for accelerated ozone depletion is growing.

Methane, as we have already seen, is a greenhouse gas which occurs from a variety of sources, including tundra areas (see Fig. 5.5). As indicated in Figure 5.5,

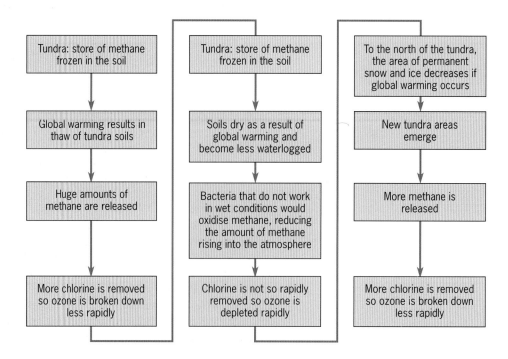

Figure 5.13 Some aspects of the relationship between methane released from tundra areas, global warming and the ozone layer

5 Essay: Assess the extent to which damage to the ozone layer is the result of human-made or natural causes.

---

this gas influences atmospheric ozone because it encourages the formation of ozone. It slows oxidation and converts ozone-depleting chlorine to hydrochloric acid; this then falls as **acid rain**. Overall, ozone may be increasing in the upper **troposphere** but it is depleting in the stratosphere (Gribbin, 1986). Figure 5.13 outlines a sequence of possible scenarios that are all linked. They concern the impact **global warming** might have on the release of methane from tundra areas and the consequent implications for ozone depletion.

Figures 5.5 and 5.13 also give an insight into the complexity of the chemistry affecting the ozone layer. Because they are so complicated, we need to use computer *models* to handle the interrelationships between the huge number of variables which influence ozone depletion.

5.4 Consequences of ozone depletion

Many people are increasingly concerned about ozone depletion because of the consequences for life on earth. Such consequences place everyone at risk – both in economically developed and economically developing countries. The risks are greatest, though, in those areas where ozone depletion is most severe.

Effects on health

Depletion of ozone could allow the penetration of wavelengths of UV radiation that normally do not reach the surface, or only in tiny amounts. In fact, some UV radiation already reaches the earth's surface – enough to give people suntans. Increasing amounts of UV could, though, lead to a number of different health problems in both people and animals (Figs 5.14–5.15).

Figure 5.14 Increased ultraviolet radiation from ozone depletion causes eye disorders such as cataracts and blindness

---

?

---

6 Construct a questionnaire to examine whether and how people's health has been affected by increasing levels of UV radiation (Figs 5.14–5.15). You should include questions about:
a age
b gender
c usual habits during periods of sunshine e.g. whether sunbathing holidays are preferred, use of sun-screens etc.
d whether people suffer/have suffered from any sun-related illness.

7a Complete the questionnaire using a variety of interviewees.
b Collate and analyse your results. You could use graphs and maps.
c Comment on your findings.

Exposure to ultraviolet radiation has been associated with damage to the cornea, lens, and retina of the eye. The principal corneal damage linked to UV exposure is photokeratitis, which appears to be related to acute UV-B exposures. The principal lenticular damage is cataract. The relationship between UV-B exposure and two forms of cataract appears to be related to cumulative exposure. It has been estimated that a 1% increase in stratospheric ozone will be accompanied by a 0.6% to 0.8% increase in cataract. Retinal damage is rare but can occur particularly in those individuals whose lens has been removed in cataract operations.

Ultra-violet radiation is known to affect the immunological defences of the skin, the first barrier of the body to foreign agents… animal models have indicated that UV-B can adversely affect the ability of animals to respond to or contain various infectious agents. Animal data suggest that an increase in the severity of certain infections may occur as UV-B fluxes increase due to ozone depletion. In areas of the world where such infections already pose a significant challenge to the public health care delivery systems, the added insult may be significant.

Figure 5.15 The impact of ozone depletion on human and animal health (*After:* UNEP Co-ordinating Committee on the Ozone Layer)

Effects on plants

Research has shown that the growth and photosynthesis of certain plants e.g. rye, sunflower and maize, can be inhibited under UV-B radiation (one part of the ultraviolet spectrum). For example, scientists have found that soya bean suffers a 25 per cent reduction in yield when exposed to a 25 per cent increase in UV-B radiation. Plants can also become more susceptible to diseases (Fig. 15.16). However, certain environmental factors like plant diseases, competition with other plants, temperature and levels of carbon dioxide (CO_2) can interact with the effect of UV-B. This makes it difficult to make quantitative predictions.

Figure 5.16 Ozone depletion and plant diseases (*Source*: Bureau of Meteorology, Australia)

?

8 Evaluate the different attitudes to ozone depletion and the dangers of CFCs in Figures 5.17–5.18.

9 Outline some possible consequences of damage to vegetation and crops in:
• Europe,
• Africa.
Give reasons for your answers.

It has also been shown that certain diseases may become more severe in plants exposed to enhanced UV-B radiation... Sugar beet plants infected with *Cercospora beticola*, and receiving 6.9kJ /m^2 per day UV-B, showed large reductions in leaf chlorophyll content, and fresh and dry weight of total biomass.

In another study, three cucumber cultivars were exposed to a daily UV-B dose of 11.6kJ/m^2 UV-B in a greenhouse before and/or after infection with *Colletotrichum lagenarium* or *Cladosporium cucumerinum*, and analysed for disease development. Two of the three cultivars were disease resistant and the other was disease susceptible. Pre-infection treatment with UV-B radiation led to greater disease development in the susceptible cultivar and in one of the disease resistant cultivars. Post-infection treatment did not alter disease development.

To date, most research has been concerned with plants from temperate regions. However, data also shows that some tropical plants may be adversely affected by an increase in UV-B radiation. Clearly, there is a need for further research and concern over the effects of UV-B radiation on agricultural systems, forestry as well as natural vegetation.

If ozone depletion lets more UV radiation reach the ground this would be harmful – but how harmful? After all, UV radiation also has some beneficial effects: for one thing it stimulates production of vitamin D in humans.

We must distinguish carefully between effects produced by UV radiation and the significance of those effects. Some wavelengths of UV light can break DNA molecules. Unlike X-rays, however, UV does not penetrate far... We should not under-estimate the capacity of organisms to endure what they cannot cure.

The idea that UV light is necessarily harmful derives from theories about the evolution of the Earth's atmosphere and about early life. The suggestion is that photosynthesising organisms evolved, that the oxygen they released led eventually to the formation of the ozone layer, and that it was only after the ozone layer had formed, and harmful UV radiation could not reach the surface, that organisms were able to colonise dry land and more advanced forms of life could appear. Although apparently attractive, the scenario leads to difficulties. If the original photosynthesising organisms had to be protected from UV radiation they must have been confined to a layer of the oceans below the level to which UV light could penetrate but shallow enough for visible light to reach them. Could photosynthesis have evolved under such dim auspices?

M Allaby (freelance writer)
Professor J Lovelock (private researcher and writer)

Figure 5.17 The pros and cons of UV radiation (*Source*: Bureau of Meteorology, Australia)

What evidence is there at ground level that UV levels have changed in response to ozone loss?

Reliable measurements of the particular form of UV radiation that causes health problems have recently been initiated in Australia and Antarctica. In both locations there is clear evidence, on an episodic basis, that reduced ozone results in enhanced UV. The natural variability of this radiation (essentially caused by clouds) probably precludes observation of long-term increases in UV corresponding to the observed ozone losses over Australia during the past decade.

According to the original Montreal Protocol, the worst case scenario for CFC-induced UV increases would be just 8 per cent, 100 years from now.

Recent calculations of skin-reddening UV increases by the year 2000, based on observed ozone losses over Australia, indicate increases in this radiation since 1980 that range from 6 per cent in Brisbane to 13 per cent in Hobart. DNA-damaging radiation increases are larger (7 per cent Brisbane) to 23 per cent (Hobart)

Dr P Fraser
(Senior Principal Research Scientist, Victoria)

Figure 5.18 Ozone: cut the hot air, here are the facts (*Source: The Australian*, 23 July 1993)

Effects on marine ecosystems

Marine phytoplankton produces almost as much biomass as all terrestrial ecosystems combined. Recent research shows that marine ecosystems are already being stressed by higher levels of UV-B as higher levels of radiation kill plankton.

Ozone deletion and global warming

Many people are confused over damage to the ozone layer and global warming. Although these dangers both involve very different processes, there are some links.

10 Draw a flow diagram to show how a reduction in quantities of plankton due to increasing levels of UV-B radiation could have wider consequences for the whole food web. Use the following processes to help you.
- Loss of plankton
- Less biomass for human consumption
- Less fish and other marine life that feed on plankton
- Increasing levels of UV-B radiation
- Birds and other creatures that eat fish suffer

For example, an increase in ultraviolet radiation reaching the surface could kill huge quantities of plankton in the oceans. We saw in section 4.3 that plankton act as a huge *sink* for CO_2. It follows, therefore, that a loss of plankton could contribute to global warming.

We have also seen (section 4.3) that ozone occurs in the troposphere (*tropospheric ozone* accounts for about one tenth of all ozone in the atmosphere). Tropospheric ozone can absorb UV radiation and it is therefore a greenhouse gas which makes a contribution to global warming. The origin of tropospheric ozone, though, is different from stratospheric ozone. Sources include intrusions of stratospheric ozone and photochemical production from nitrogen oxides, hydrocarbons and carbon monoxide from transport and industry. Ozone in the troposphere rises very slowly into the stratosphere where it can contribute to ozone depletion.

Other impacts

UV-B radiation has also been shown to degrade wood and plastic products, leading to discoloration and a loss in strength. There is still, of course, a need for more data on the way different materials respond to increased levels of radiation in different parts of the world, but UV-B radiation is known to break down polymers used in buildings, paints, packaging and other products.

Changes in the temperature of the ozone layer due to different amounts of ultraviolet energy being absorbed could also affect air movements in the stratosphere. This may be particularly significant over the equator where the ozone layer is currently at its thickest. Similarly, this could have as yet unknown consequences for the world's wind patterns and even oceanic circulation. In turn, this would influence the world's **weather** and **climate**.

The response in Australia

Average ozone decreases over Australia (Fig. 5.19) have amounted to 0–6 per cent between 1980 and 1990, with the higher losses occurring in the more southern regions. Subsequently, Australians have become increasingly concerned about the damage to the ozone layer. Table 5.2 shows spatial variations in the level of response to a question asking Australians which environmental problems they were most concerned about. Of the global issues listed, ozone depletion rated as being of more concern (29 per cent) than global warming (17 per cent) and overpopulation (13 per cent).

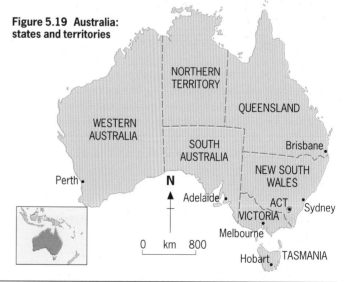

Figure 5.19 Australia: states and territories

Table 5.2 Variations in levels of concern over ozone depletion between Australian States and Territories, based on a survey carried out in May 1992 (*Source: Environmental Issues*, Australian Bureau of Statistics, 1993)

State or territory	New South Wales	Victoria	Queensland	South Australia	Western Australia	Tasmania	Northern Territory	ACT (Australian Capital Territory)	Australia
Percentage	27.5	31.8	26.2	29.1	25.7	28.1	38.5	34.3	28.6
Numbers in survey (millions)	1168.9	1028.1	551.5	312.9	305.1	92.6	36.0	70.7	5767.8
Total population (millions)	4243.8	11028.1	2102.9	1075.6	1188.9	329.1	93.6	206.0	20268.0

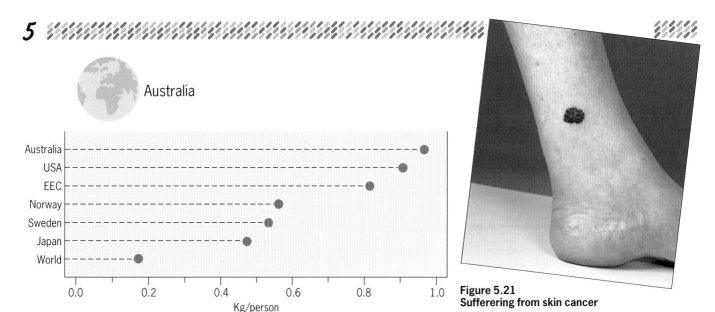

Australia

Figure 5.20 Per capita consumption of CFCs for selected countries, 1985–6

Figure 5.21
Sufferering from skin cancer

Figure 5.22 Slip on a T-shirt, slop on the sunblock

Get **SunSmart**

SPF 15+

Figure 5.23 SunSmart campaign

Australia was a major per capita user of CFCs and halons (Fig. 5.20) in the 1980s. The largest consuming uses were refrigeration and air conditioning. The third major source in 1988 was the production of aerosols. The Australian government and industry have taken many steps to reduce CFC consumption since 1986. For example, Australia passed the Ozone Protection Act in 1989 that imposed a total ban on the manufacture and importation of aerosols containing CFCs from the 31 December 1989. The use of aerosols was restricted to some medical applications and was reduced from about 4.5 thousand tonnes to 300 tonnes in 1991. A total phaseout of CFCs is planned by 1997.

Australia's warm climate means outdoor activities are the nation's most popular recreational pastimes. One impact of the changes to the ozone layer is the increased risk of sunburn and associated skin cancer (Fig. 5.21). Australia has the highest rate of skin cancer in the world with two out of three Australians developing some form of skin cancer during their lifetime. The growing understanding of the dangers of severe sunburn has resulted in campaigns to increase public awareness of the

problem (Figs 5.22–5.23). In fact, Australians have successfully adapted their lifestyles in other ways too, e.g. schools avoid outside activities during the middle of the day when **insolation** and ultraviolet radiation are at their strongest.

11 Why are Australia and New Zealand particularly affected by ozone depletion?

12a Use Table 5.2 and Figure 5.19 to draw a choropleth map of Australia.
b Try to account for the spatial variations shown by your map in levels of concern over damage to the ozone layer.

13 Study Figure 5.20 and suggest why the consumption of CFCs was particularly high in Australia in the 1980s.

14a What protective measures are suggested in Figures 5.22–5.23?
b Design your own publicity and information programme to increase the Australian public's awareness of ozone depletion. Include reasons for the warnings you give.

15a Consider the attempts to reduce the problem of ozone depletion.
b Suggest what other steps
• we as individuals, and
• manufacturers could take.

16a Select a variety of recently developed products which aim at the 'ozone-friendly' market.
b In your opinion, how successful are their advertising/sales?
c Assess the products according to whether you think they meet their claims.
d Summarise your own response to manufacturers' attempts to combat ozone depletion.

5.5 Attempted solutions to the problem

Removing CFCs

If CFCs are the major culprit of ozone depletion, an obvious response is for governments to ban them. This happened in many economically developed countries after the 1987 international agreement at Montreal (see Table 5.3). For example, the USA banned the use of CFCs in aerosols in 1987. Many countries, though, did not follow such policies – particularly those from the economically developing world.

Since 1987, manufacturers responded to the ozone problem by trying to develop replacements for CFCs that would not harm the ozone layer. In some cases the solutions were relatively simple, such as many fast-food chains changing their packaging to cardboard boxes, or transferring to the less harmful CFC 22 which is used in some food packaging and also in most industrial refrigeration. Similarly, many aerosol products have been replaced by roll-ons and sticks or use a pump action spray rather than a propellant gas. In fact, as you read, you may be praising yourself for purchasing so called 'ozone friendly products', and perhaps we should all play our part in protecting the ozone layer through our purchasing behaviour. However, we need to be aware that products claiming to be ozone friendly often contain chemicals other than CFCs that can still be damaging.

Other chemicals

Considerable effort and expense has also been involved in the development of complex alternatives to CFCs. HFCs and HCFCs (hydrofluorocarbons and hydrochlorofluorocarbons) are less damaging than CFCs but not totally harmless, so their use needs to be carefully controlled. One problem is that the development costs (research, labour and materials) encourage a long period of use so that companies can recover their investment. This, of course, is not in the interests of the ozone layer. In addition, many chemical companies are unwilling to share their technology with economically developing countries unless they can profit by doing so.

Global agreements

Ozone depletion is clearly a global problem requiring global solutions. International efforts are therefore needed to combat the growing problem. These began with the agreements of the Montreal Protocol in 1987 (Table 5.3).

Table 5.3 International steps that have been taken to protect the ozone layer (*After: The Ozone Layer*, DoE)

Early 1980s Independent action by some countries to control ozone depletion. Realisation that global action needed.

1985 Vienna Convention for the Protection of the Ozone Layer agreed.

1987 Montreal Protocol: required production of CFCs and the most damaging chemicals to be frozen at 1986 levels and to be halved by 1999 (150 countries needed to sign the convention for it to be effective).

1989 Thirty countries signed the Montreal Protocol.

March, 1990 Saving the Ozone Layer conference, London.

May, 1990 First meeting of the Parties to the Montreal Protocol. Helsinki Declaration adopted for CFCs and halons to be phased out by 2000.

June, 1990 Second meeting of the Parties to the Montreal Protocol in London. Agreement to phase out CFCs, halons, carbon tetrachloride by 2000 and trichlorethene by 2005. Spurred on by subsequent reports that the ozone holes had been getting worse, sixty countries signed the Montreal Protocol. The international 'Montreal Fund' established to help pay for the phasing out of ozone-depleting chemicals in the developing world and to pay for the transfer of technology.

1991 Alarm in many economically developed countries as the media brought the problem to the public's attention. Some countries consequently pressed for bans on CFCs much earlier: Germany and Denmark pledged to ban the use of CFCs by 1995; the USA by 1995; the EC by 1997 (three years ahead of the date set by the Montreal Protocol).
One difficult area: economically developing countries, like India, that use CFCs heavily in air conditioning and refrigeration and are less able to afford expensive substitutes.

June 1992 Second meeting of the Parties to the Vienna Convention and third meeting of the Parties to the Montreal Protocol.
Fourth meeting of the Parties to the Montreal Protocol. The steps taken represented progress but not to the degree that was hoped; a total ban on the worst CFCs was brought forward from 2000 to 1996.

December, 1992 The EC countries made their own agreement to phase CFCs by 1995. Controls on other ozone-depleting chemicals were also announced, e.g. halons to be phased out by 1994 rather than 2000. Funding of £370 million agreed to finance assistance to less developed countries. (This has not always been backed by payments. As a number of countries have not lived up to their commitment.) A complex timetable was also announced for HFCs, rather than a total ban. This included a 65 per cent reduction by 2010 and 100 per cent by 2030. Although HFCs are less damaging than CFCs, they act faster. The adopted timetable means that HFC damage will peak at about the time the hole in the ozone layer reaches its maximum from the longer-lived CFCs.

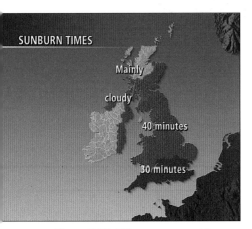

Figure 5.24 TV announcement for sun exposure

Some countries have taken steps to reduce ozone depletion, but scientists estimate that even if CFCs were phased out totally, it would take another 80–90 years before the ozone layer was restored. The timetables adopted at the Copenhagen meeting in 1992 mean that this recovery will take longer and therefore expose people to unnecessary risks. There is no doubt that the situation will get worse before it gets better. In fact, for the first time in the summer of 1994, television weather forecasts also included maps indicating how long we could safely stay in the sun before experiencing sunburn (Fig. 5.24).

In countries like Australia and Chile that are already affected by the Antarctic hole, there has been considerable complacency over the damaging effects of UV radiation. In contrast, although in Europe we have time to prepare should the Arctic hole get any larger, skin cancers often do not develop until several years after a period of exposure to high levels of radiation. It is therefore necessary to carry out research, particularly in the southern hemisphere, to establish as accurately as possible the worst effects of higher levels of UV radiation.

SEVENTEEN countries, including the US, agreed last week to cut their emissions of the ozone-destroying chemical methyl bromide by a quarter by the year 2000.

But at the same time, US President Bill Clinton was promising congressmen there would be no restrictions on methyl bromide before 2000, in return for their support for the North American Free Trade Agreement.

In Bangkok last week, Elizabeth Dowdeswell, the head of the UN Environment Programme, told the annual meeting of signatories to the Montreal Protocol that 'the state of depletion of the ozone layer continues to be alarming', with record lows in both hemispheres this year.

Scientists agree that methyl bromide, a crop fumigant, has caused between 5 and 10 per cent of the destruction of the ozone layer. Last week the 17 countries agreed not only to cut methyl bromide use by 25 per cent by 2000, but also to phase it out 'as soon as technically possible'.

Earlier this month Mickey Kantor, the US's international trade representative, wrote to

Florida Farmers promising that 'there will not be any restriction on the manufacture and use' of methyl bromide 'until the year 2000'. Yet the US had already agreed to freeze production at 1991 levels by 1995. His letter was part of the lobbying to secure support for the NAFTA. Four Florida representatives, from districts where growers oppose limits on methyl bromide, voted for NAFTA, which Congress passed by 234 to 200 votes.

Jacques Rosas of Greenpeace in Washington DC says Clinton's promises also secured votes from representatives in his home state of Arkansas. Ethyl Corporation of Magnolia and Great Lakes Chemical Corporation of El Dorado, both in Arkansas, make all the US's methyl bromide – 43 per cent of the world's consumption. All Arkansas senators and representatives voted for NAFTA.

It also emerged last week that agreements to limit ozone-destroying chemicals may have little effect in developing countries. Dowdeswell said in Bangkok that industrialised countries have cut emissions of ozone-destroying chemicals by 45

per cent, but only nine developing countries have cut emissions.

Partly as a result of these statistics, the parties to the protocol voted to double the Interim Multilateral Fund, set up in 1991 to help developing countries phase out ozone-destroying chemicals. British complaints that only $150 million of the original $240 million has been spent were met by allowing Britain to contribute to the new fund in 'promissory notes', to be cashed only when existing funds are used up.

Bill Walsh of Greenpeace says 60 per cent of projects supported by the fund include CFC replacements which themselves pose problems. Walsh cites a grant of $8.3 million to the UN Industrial Development Organization to retool five refrigerator manufacturing plants in Iran, so that they use HCFCs as coolants.

HCFCs are safer than CFCs, but still destroy some ozone. Industrialised countries have pledged to ban HCFCs by 2030, and in Bangkok twelve European countries pledged to ban them by 2015, but developing countries have not accepted any restrictions on HCFCs.

Figure 5.25 Clinton faces both ways on ozone treaty (*Source: New Scientist*, 27 Nov. 1993)

?

17a What does Figure 5.25 suggest is a major obstacle to controlling ozone depletion?
b Explain why such obstacles make it difficult to phase out ozone-depleting products.

18 Essay: Discuss the importance of international initiatives like the Montreal Protocol in controlling ozone depletion. What are the main obstacles to such efforts?

19 Our understanding of ozone depletion encourages the precautionary principle. Do you agree? Write an article on ozone depletion. Set out and justify your opinion on the role of individuals and governments in tackling this problem.

Summary

- The ozone layer filters out ultraviolet radiation from the sun.
- A hole in the ozone layer over Antarctica was discovered by British scientists in 1985. A similar hole was detected by satellites over the Arctic in the early 1990s.
- Ozone naturally breaks down in the stratosphere but it also naturally replenishes itself. Pollution accelerates the rate of ozone destruction to the extent that natural replacement is too slow to compensate for the damage.
- The principal cause of ozone depletion is human-made and involves chemicals like CFCs. There are also a number of natural causes including volcanic eruptions.
- Ultraviolet light can cause skin cancer and eye problems for people and damage vegetation and crops. Without the ozone layer, life as we know it would not exist on earth.
- Ozone is a greenhouse gas. Ozone depletion could increase global warming as CFCs are also effective greenhouse gases and influence climate change.
- Manufacturers have taken many steps to reduce the use of CFCs and other damaging chemicals. In many cases, new products have been developed which do not involve ozone-depleting chemicals.
- Damage to the ozone layer is a global problem that requires international co-operation and initiatives. The principal step so far has been the 1987 Montreal Protocol.

6 Atmospheric movement

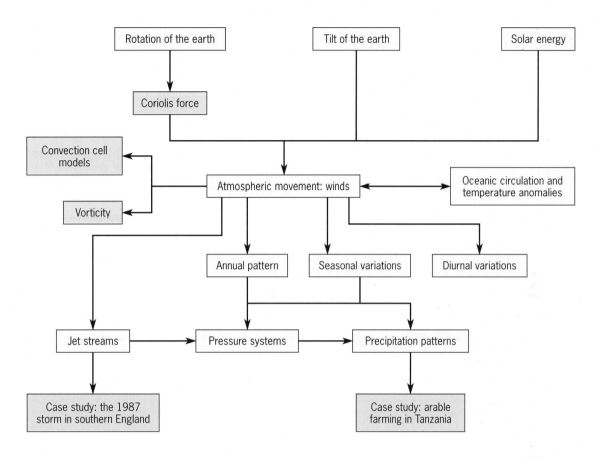

6.1 Introduction

We have previously established that the energy that drives the earth's **climate** and **weather** patterns comes from the sun (section 2.5). The earth's surface absorbs this energy largely in the tropics and near the equator (see Figs 2.4 and 2.5). It is this uneven distribution that is largely responsible for motion in the atmosphere and oceans. Circulations of air and water redistribute heat energy from hotter areas where there is an energy surplus, to colder areas with a deficit (see section 2.5). The resulting patterns of winds and ocean currents are very complex and also difficult to explain. The circulation of the atmosphere also affects and results from the distribution of the world's pressure belts (see Figs 6.8 and 6.9).

Although the ocean currents tend to be more stable in their position, we shall see that there are important links between atmospheric circulation and the 'weather and climate' of the oceans. As stated in Chapter 1, global circulation models involving both the atmosphere and oceans are now also fundamental to computer-based weather forecasting.

Collectively, the patterns of winds, ocean currents and pressure systems determine the world's weather and climate patterns and consequently influence human behaviour. For example, seasonal changes in winds and **pressure** systems strongly affect agriculture, particularly in the tropics (Fig. 6.1). We will, in fact, consider the extent to which arable farming patterns are the result of seasonal changes in the weather, or other factors.

Figure 6.1 Arable farming; planting cassava in Tanzania to coincide with the beginning of the wet season, early April

6.2 Winds

Movement of air in the atmosphere results from the uneven distribution of solar energy. This is because the uneven heating of the earth's surface causes variations in air pressure that in turn produce air movement or wind. **Isobars** show these pressure systems on synoptic weather charts (Fig. 6.2). An isobar is a line that joins points of equal atmospheric pressure, just as contours that show relief on land. In fact, meteorologists refer to **ridges** of high pressure, while a **trough** is an elongated area of low pressure. Continuing with this analogy, the rate of pressure change is indicated by the spacing of the isobars, called the **pressure gradient.** This is one of a number of forces that influence the direction and strength of winds.

Pressure-gradient force
Just as water flows down a slope, it follows that if there is a pressure gradient force brought about by spatial differences in heating, then air will move down the gradient from high to low pressure (Fig. 6.3). In other words, air tends to move horizontally when there is a variation in pressure at the same altitude.

1a Draw a copy of Figure 6.2 and label • a steep pressure gradient and • a gentle pressure gradient.
b On your map indicate how a steep pressure gradient and a gentle pressure gradient would relate to wind speed.

2 Compare Figure 6.2 and Figure 6.4. Which letters on Figure 6.4 correspond to the following:
• an area of high pressure
• an area of low pressure
• a ridge
• a trough?

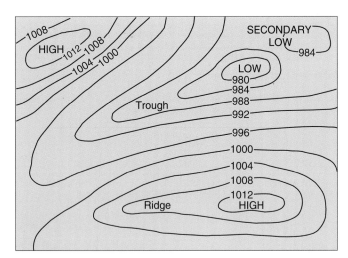

Figure 6.2 Terminology associated with isobaric maps (above)

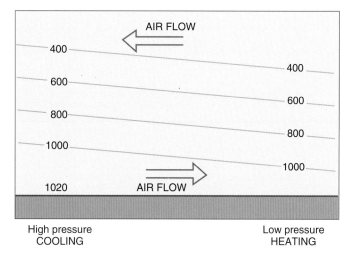

Figure 6.3 The pressure-gradient force resulting from differential heating (above right). This simple diagram only applies at a local scale; on a larger scale air flow is affected by other factors.

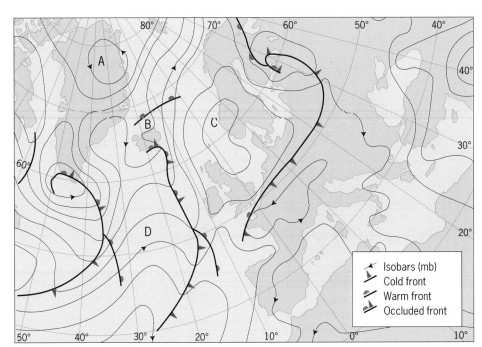

Figure 6.4 Weather chart for the North Atlantic and northern Europe, 18 July 1994 (*Source:* Met. Office)

Coriolis force

Every point on the earth's surface rotates once in a period of 24 hours. A point near the poles traces a tiny circle and moves very slowly (Fig. 6.5). At the equator, a point has to travel a distance of a little over 40 000 km in the same 24 hours – it does this at a speed of 470 m/s. Consequently, air moving polewards carries the same momentum that it started with at the equator. However, further north or south, this momentum is too great for the latitudes. The result is the deflection of winds to the east in the northern hemisphere and to the west in the southern hemisphere. The **Coriolis force** is the term for this influence on winds.

In the mid-latitudes and in the mid-**troposphere** (see Fig. 9.1) the pressure-gradient force and the deflecting effect of the Coriolis force are in balance. This leads to air moving not from high to low pressure but between the two, parallel to the isobars, called a **geostrophic wind** (Fig. 6.6). If we now introduce curved isobars, a *centrifugal force* comes into play. This is a tendency for the air, as it rotates around the system, to move into lows or out of highs. A force called the *centripetal force* related to the acceleration of the wind is required to maintain a flow parallel to the isobars and produce a **gradient wind.**

Nearer the surface, another variable is involved which is the influence of friction with the ground (Fig. 6.7). This reduces the deflecting influence of the Coriolis force and increases frictional drag which in turn also reduces wind speed. This causes air to turn across the isobars and, in the case of a low pressure system, this is at an angle of about 15° over the oceans and 30° over the land.

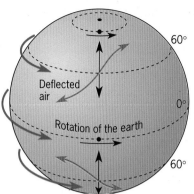

Figure 6.5 Deflection of winds by the Coriolis force

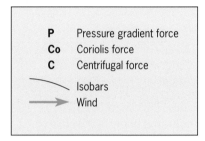

Figure 6.6 How winds are steered in the upper atmosphere

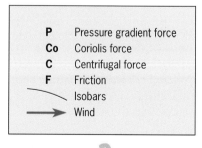

?

3 Look at Figure 6.14. How do the cloud patterns in this view of the earth reflect the influence of the Coriolis force?

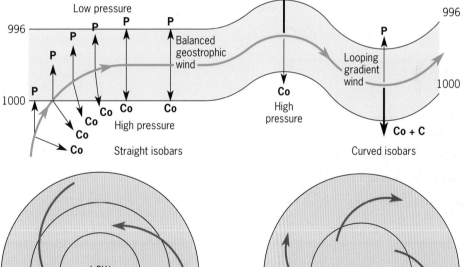

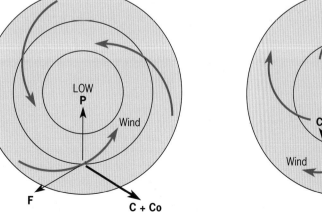

Figure 6.7 How winds are steered near the ground in low and high pressure systems

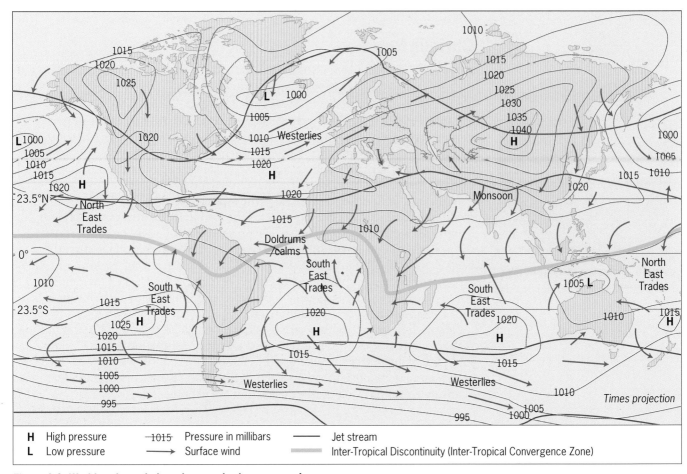

Figure 6.8 World surface winds and atmospheric pressure, January

| H | High pressure | ──1015── | Pressure in millibars | ──── | Jet stream |
| L | Low pressure | ──→ | Surface wind | ▓▓▓▓ | Inter-Tropical Discontinuity (Inter-Tropical Convergence Zone) |

Global air movements

Air movements or winds can occur at a variety of scales. This chapter will concentrate on the large-scale planetary motions, including major winds like the Trade winds that exist largely in the tropics. There is a close relationship between major winds and the world's major pressure systems (Figs 6.8–6.9). There are also seasonal variations that we will examine in section 6.4, the principal change being in south Asia where there is a reversal of pressure between January and July leading to the **monsoon.**

The large global wind belts also contain major weather systems, such as **hurricanes** in the tropics. In the mid-latitudes **cyclones**, or **depressions**, can be up to 1500 km in diameter and have an important effect on the weather in these parts of the world. There are also small-scale local winds related to localised heating and pressure differences that are often influenced by relief. Tornadoes or whirlwinds (waterspouts over the sea) occur at an even smaller scale (Fig. 6.10).

Figure 6.10 Tornado

Tornado A tornado is an intense low pressure system with a small diameter often of less than 500 m. They are caused by intense heating and fed by warm air currents converging on the low pressure. Tornados are areas of strong upward air currents and powerful circling winds.

Tornados can cause immense damage to structures and vegetation over which they pass. There have been many unusual accounts of people and animals being lifted and carried some distance before being dropped – still alive!

Over the sea they form waterspouts, lifting sea water into the air plus any life in the water. This may account for rare events like fish and frogs falling from the sky when it rains!

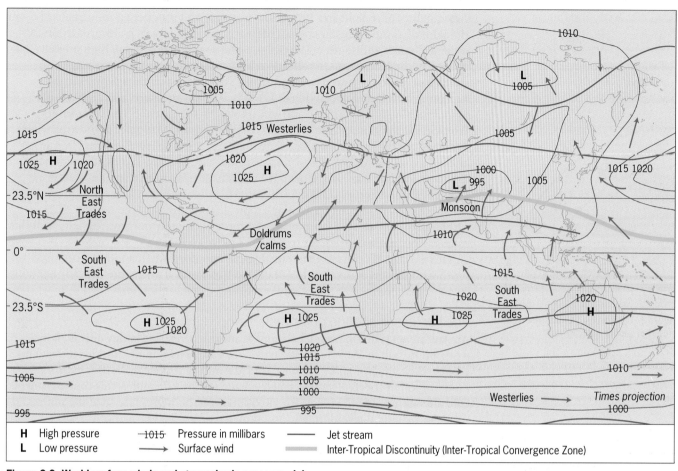

Figure 6.9 World surface winds and atmospheric pressure, July

| H | High pressure | ——1015 | Pressure in millibars | —— | Jet stream |
| L | Low pressure | ——→ | Surface wind | ▬▬▬ | Inter-Tropical Discontinuity (Inter-Tropical Convergence Zone) |

6.3 Global circulation

The global movements of winds largely result from the differences in the amount of solar radiation received in the tropics and the polar latitudes. Other factors are also involved such as the position of the continents, major relief barriers like the Himalayas and the differences between the degree of heating of the atmosphere over land and sea areas.

Convection cell models

We have seen that low latitudes are warmer because they have surplus energy, whereas the high latitudes are colder having an energy deficit (see section 2.4). We should expect this to result in a large **convection** cell with air rising over the equator due to strong heating. This air would then move polewards to sink and be drawn back to the low pressure over the equator (Fig. 6.11). However, this simple model does not take account of other factors, particularly the spinning of the earth. In fact, for many centuries sailors have learnt about the world's winds as they navigated in different oceans. They found that there are belts of winds on the planet, some moving towards the equator with others moving in the opposite direction.

The Hadley cell

The 18th century meteorologist George Hadley was the first to develop a more complex model in 1735. We still use the **Hadley cell** today to explain atmospheric circulation in the tropics, and it emphasises the importance Hadley gave to the

Figure 6.11 Simple convection cell on a non-rotating earth

81

Convection cell models

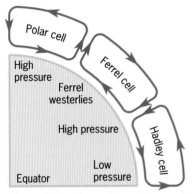

Figure 6.12 The traditional three cell model of global atmospheric circulation

---

?

---

4 Study Figure 6.13. Why do the Trade winds not approach the equator at right angles as they are drawn in to the equatorial zone of low pressure?

5 At approximately what latitude does a jet stream occur in the sub-tropics (Figs 6.8–6.9)?

tropics as a heat source. Intense **insolation** causes warm air to rise by convection at the equator. This results in a belt of low pressure and produces a zone of cloud cover and rainfall as the rising air cools. This warm air then travels polewards to sink at about 30° north and south of the equator and so produce areas of high pressure. As the descending air warms, its ability to hold moisture increases resulting in an absence of cloud cover and **precipitation.** This sinking air then returns to the equator to complete the cell (Fig. 6.12). A belt of low pressure, also called the *equatorial trough*, draws the Trade winds towards the equator. The term **Inter-Tropical Convergence Zone (ITCZ)** is used by most texts to refer to the area where the Trade winds meet. However, many meteorologists dislike the term ITCZ as it implies a continuous belt of low pressure and rain, as if the Trades converge. As a result, alternative terms have been used like heat trough, heat low and **Inter-Tropical Discontinuity (ITD)**. As this area is a discontinuous zone, it is more suitably described by the term Inter-Tropical Discontinuity (see section 10.2). Although we will use the latter term as it more accurately portrays the nature of this zone, many texts will still refer to the ITCZ.

Note that the ITD forms a **meteorological equator,** one that moves north and south with the changing position of the overhead sun. This section already explains some of the general features of equatorial and tropical climates.

The Ferrel cell and the polar cell

An American, William Ferrel, had similar ideas to those of Hadley although his work, published in 1889, focused on cyclones and winds in the mid-latitudes. This pioneering meteorologist gave his name to a mid-latitude cell. In the traditional model, another thermally related cell existed towards the poles, this time driven by cold air sinking over the poles. Areas of persistent high pressure sit at the poles and winds move equatorwards to rise and return polewards (Fig. 6.12).

New circulation models

We have learnt much more about global circulation in recent years, particularly from satellites, that has led to new ideas and models. The basic model of atmospheric circulation today (Fig. 6.13) is based on research by Palmen (1951) and little has since changed in its applicability at the global scale. The latest model still retains the three convection cells in each hemisphere, but their relative importance and roles have changed. Figure 6.13 shows that the troposphere thins towards the poles and that the **tropopause** is also broken into three distinct sections. It is at these breaks that powerful eastward-moving high-altitude winds called **jet streams** occur (see section 6.5).

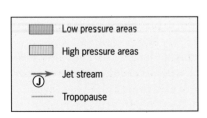

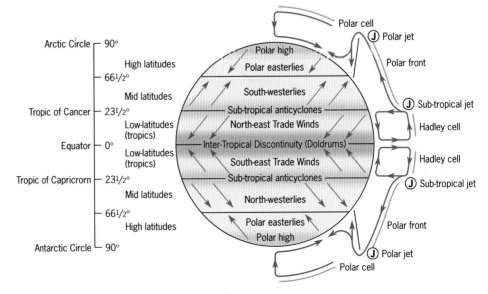

Figure 6.13 Global atmospheric circulation and major pressure belts

The Hadley cell still forms the basis of understanding atmospheric circulation in the tropics. However, in the 1970s Walker made a modification to the tropical circulation model held for so long. He identified a cell that operated on an east-west axis rather than north–south. Figure 6.15 shows that it also begins with warm air rising by convection near the equator so that towering *cumulonimbus* cloud is produced. This air then moves with the sub-tropical easterly jet stream before descending. Further research is needed to establish more clearly the importance of these cells to climate and weather in different parts of the tropics.

Figure 6.14 Cloud patterns over the earth, January 1982

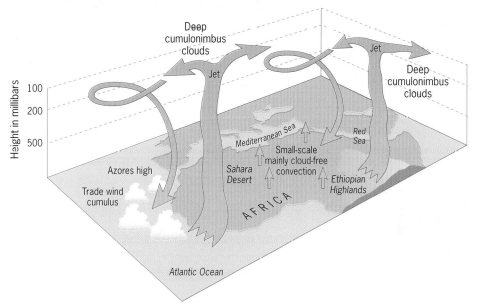

Figure 6.15 Air circulation in the Tropics (*After:* Walker, 1972)

Mid-latitude movements

Between the tropical and polar circulations there is a more complex zone. Early in the twentieth century, research into patterns of winds and clouds in the upper troposphere (3–12 km) revealed belts of large-scale, westerly winds (Figs 6.8, 6.9 and 6.13) called the **Ferrel westerlies**. These winds form a powerful flow of air that travels around the pole, so we sometimes refer to it as the **circumpolar vortex**. The upper-altitude westerlies have an important effect on surface winds and pressure systems, and it is here, within these belts that the separate, fast-moving jet streams exist.

In the mid-latitudes, the Coriolis force that deflects winds is in balance with the pressure gradient force. Winds consequently move along the isobars and are known as geostrophic winds. Figure 6.16 shows isobars in the upper atmosphere and those which are close together indicate the position of the jet stream and the polar **front**.

These planetary westerly winds occur in a series of waves known as the **Rossby waves**. There are usually between two and five in the northern hemisphere and three south of the equator. The causes of these waves are beyond the scope of this book and involve very advanced physics. At this point though, we can say that they are a response to, firstly, the thermal differences between the equator and the poles and, secondly, the rotation of the earth. Major relief barriers, like the Rockies and Himalayas in the northern hemisphere and the Andes in the south, may also influence the stability of these waves. In addition, the Tibetan Plateau could have a deflecting effect on the waves in the northern hemisphere.

The westward-moving depressions and **anticyclones** (not to be confused with the largely stationary pressure systems referred to previously) are a surface result of

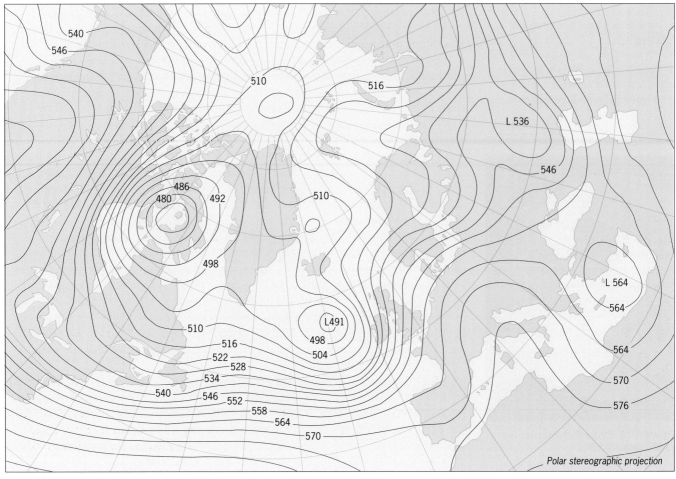

Figure 6.16 Upper atmosphere isobars across the northern hemisphere (Source: Met. Office)

these high-altitude waves. Consequently, horizontal movements of energy are now more important than the idea of a mid-latitude convection cell. The formation of waves within these air movements and their links with near surface pressure systems have much to do with the concept of **vorticity**.

Vorticity

One definition of vorticity is the amount of spin possessed by a rotating body (Wright, 1983). An ice skater has vorticity when s/he spins. This is a useful analogy for the way in which the rotation of the earth encourages air to rotate. Figure 6.17 shows that, because the linear movement of the earth at the poles is zero, the vorticity of air is at its maximum. The same effect occurs when an ice skater draws their arms in towards their body. At the equator, linear motion is at its maximum and a point on the surface of the earth moves parallel to the axis of rotation. This reduces vorticity to zero. It follows that points between the equator and the poles experience a combination of linear motion and vorticity. In addition, remember that we have already seen a number of other forces which also influence air movements, such as friction, pressure and the Coriolis force.

Figure 6.18 illustrates what happens to air as it moves towards a relief barrier like the Rockies. Initially, vorticity is high but as the air ascends over the mountains compression takes place and the air diverges or spreads outwards. This is like the ice skater who crouches so that their vorticity and rate of spin decreases. As the air

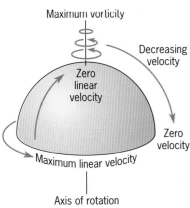

Figure 6.17 Vorticity and linear air movements on a rotating earth

Figure 6.18 Relief barriers and vorticity

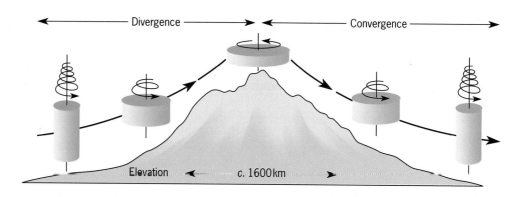

7a What happens to the rate of spin of ice skaters when:
• they stand and bring their arms towards their body,
• when they crouch towards the surface of the ice?
b Explain the link between the behaviour of an ice skater when spinning and the pattern of winds in Figure 6.13.

8 Outline three climatic effects of the movement of air within the Hadley cell shown in Figure 6.13.

9 Essay: Describe and explain the patterns of surface pressure and winds shown in Figure 6.9 in the northern hemisphere.

descends on the leeward side of mountains, the reverse occurs. Here, while the air expands vertically, vorticity increases and air converges. This means that towards the crest of mountains there is an increase in anticyclonic tendencies, while cyclonic tendencies take place on the leeward sides. There is then an effect on surface pressure and air movements (Fig. 6.19).

The westerly upper air currents wave both in a horizontal and vertical plane (Fig. 6.19). Air accelerates around the tighter bends of the ridges and then slows towards the trough. Where the upper air slows and the pressure is lower, convergence occurs. Air from high in the troposphere tends to sink, resulting in high-pressure conditions near the surface and diverging air. The opposite occurs where the westerlies accelerate towards the ridges. Here, air diverges and results in surface convergence and low pressure.

Just as the idea of a mid-latitude cell has been dismissed, there is now also doubt over the existence of a polar cell. In addition, the polar high pressure areas are no longer thought to be stable features. However, cold air does flow equatorwards from the poles towards the polar front, particularly in the winter months. Figure 6.13 shows the troposphere to be thinner at these latitudes and it seems that they may not therefore be so important to the overall global circulation.

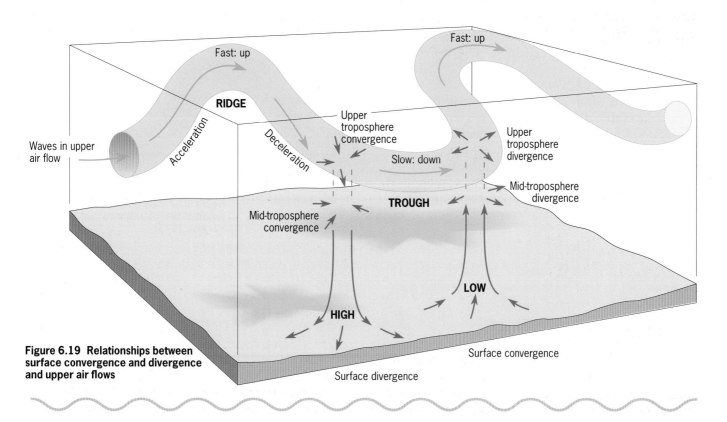

Figure 6.19 Relationships between surface convergence and divergence and upper air flows

6.4 Seasonal variations in circulation

Figures 6.8 and 6.9 show that there is very little seasonal variation in global circulation. The most noticeable change is the movement of the pattern, and particularly the ITD, north and south with the seasons. This is the result of the changing position of the earth on its axis relative to the sun and the consequent movement of the overhead sun between the two tropics. The Rossby waves, resulting from these changes, will also move and subsequently affect the climate of the British Isles and Europe (see sections 9.3 and 9.5). Although the maps do not show it, the jet streams also migrate both from year to year and on the scale of decades; the impact this can have on weather is considered in section 6.5. The one clear exception is the very pronounced changes in pressure over southern Asia that affect the monsoon (see section 10.4).

Seasonal variations in pressure and the consequences for patterns of rainfall can exert a powerful influence on human activities. A useful illustration is arable farming practices in East Africa and their adaptations to the pronounced wet and dry seasons of the tropical (savanna) climate (see Fig. 6.1).

The effect of seasonal changes in circulation on arable farming in Tanzania

Studies of the climate in east Africa generally emphasise water availability as the main limiting factor in crop production. Although this is an important influence, other factors may also be involved.

Tanzania's climate reflects its position just south of the equator. Typically, it has a pronounced wet and dry season with rainfall largely restricted to five or six months in a year over most of the country (Fig. 6.20). The principal cause of this pattern is the convergence of the Trade winds at the equatorial trough (ITD), and it is the ITD's movement which causes the wet and dry seasons (Fig. 6.21).

The significance of evaporation

The significance of any rainfall has to be considered against the **evaporation** taking place. As a result of Tanzania's tropical location, **potential evapotranspiration** rates (the maximum possible amount at a given temperature, see section 7.3) are high throughout the country (Fig. 6.21). Obviously, rates of evaporation are at their lowest during the wet season and they generally also decrease with altitude. This is because of the lower air

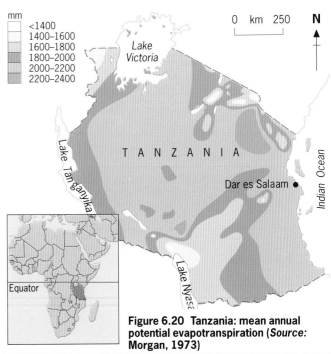

Figure 6.20 Tanzania: mean annual potential evapotranspiration (*Source:* Morgan, 1973)

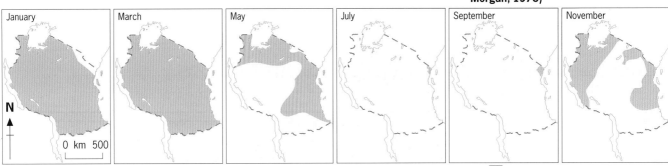

Figure 6.21 Tanzania: monthly rainfall (*Source:* Morgan, 1973)

Mean monthly rainfall of 5mm or more

temperatures that in turn reduce the ability of the air to take up moisture. For successful agriculture, therefore, maps showing spatial variations in rainfall and evaporation rates are crucial (Fig. 6.22).

Farming practices and the climate

In Tanzania, patterns of agriculture largely reflect spatial variations in the total amounts of rainfall and amounts lost through evaporation (Fig. 6.23). The occasional convectional storms that occur outside the wet season do not have any significant bearing on arable farming. Tanzanian farmers therefore grow crops adapted to shortages of moisture e.g. maize, millet and sorghum. The actual timing of the planting and harvesting of crops relates to the occurrence of the wet and dry seasons. In the wetter part of east central Tanzania, for example, two overlapping sowings of crops are common. In particular, farmers sow maize towards the end of the year to catch the rains; they harvest this in January and February. A second sowing of crops like cassava and rice is then timed to catch

10 Explain the changing distribution of monthly rainfall over East Africa shown in Figure 6.21.

11a Use the data in Table 6.1 to compile two adjacent dispersion diagrams. Show the position of the median value and the upper and lower quartiles.
b Discuss the consequences of the distribution shown on your completed diagram for agriculture in Tabora and Songea.

12 Describe and explain the pattern of arable farming in Tanzania using all the information in this section.

13 Apart from climatic influences, what other factors might have influenced the pattern of arable farming in Figure 6.23?

the heavier rains from April to June. Research has shown that such timing is important. For example, maize planted late will produce low yields because of a lack of water as well as inadequate supplies of nutrients, particularly nitrogen.

Table 6.1 Total annual rainfall (mm) for Tabora and Songea in Tanzania, 1962–88 (*Source:* CRU University of East Anglia)

Year	Songea 10°40S, 35°40E	Tabora 5°2S, 32°57E
1962	1295	1232
1963	1405	1044
1964	1140	1033
1965	1337	929
1966	999	880
1967	1388	1300
1968	1308	893
1969	663	987
1970	1221	960
1971	1131	862
1972	1134	943
1973	1403	936
1974	1200	1246
1975	1096	870
1976	797	881
1977	1082	1141
1978	1439	1147
1979	1408	998
1980	1301	803
1981	1068	829
1982	1271	966
1983	1055	986*
1984	1183	986*
1985	922	986*
1986	1356	898
1987	944	946
1988	709	931

*Estimates based on an average over the period 1962–88

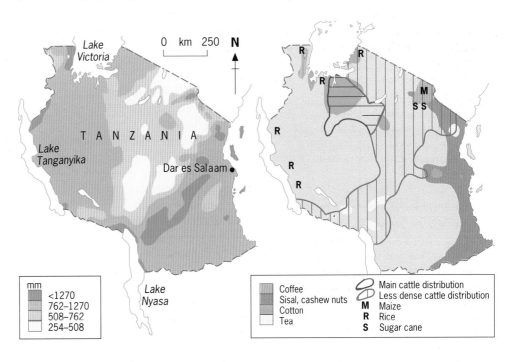

Figure 6.22 Tanzania: mean annual rainfall (*Source:* Morgan, 1973)

Lake Victoria

0 km 250 N

T A N Z A N I A

Lake Tanganyika

Dar es Salaam

Lake Nyasa

mm
<1270
762–1270
508–762
254–508

Figure 6.23 Tanzania: main commercial crops and livestock

R

R

R

M
S S

R

R

R

Coffee
Sisal, cashew nuts
Cotton
Tea

Main cattle distribution
Less dense cattle distribution
M Maize
R Rice
S Sugar cane

6.5 Jet streams

The pilots of jet aircraft were the first to became aware of the existence of jet streams, hence their name (see section 1.2). These high-altitude (about 12 000 m) winds can reach speeds of over 200 km/h. There are three main jet streams that occur at the boundaries between the three major belts in the global atmospheric circulation (see Fig. 6.13). These are areas where warm and cold air mix at high altitudes. They are therefore important locations for the transfer of energy in the atmosphere. The Polar Front Jet Stream lies at about 30–50° north

Flying the snow across the sea

The jet stream over the Atlantic may be responsible for the recent run of hot British summers – and an exceptionally cold October

What is happening to the climate in Britain? The Eighties was the warmest decade this century and had the worst drought in 300 years, global warming was supposed to be taking a grip, and the outlook was even hotter. But now we have endured a wretched summer and one of the coldest Octobers this century. The fact is that climatic change is still much more complicated than usually portrayed. Global warming may still be occurring, but many other long-term factors also affect the weather. One of these is the jet stream, which meteorologists believe may be one of the main causes of this year's sudden deterioration…

There are already chilling omens for this winter. Last month was one of the coldest Octobers this century, more than eight degrees below the average temperature. Sub-zero temperatures hit most of the country, with snowstorms sweeping Scotland and northern England. It was no coincidence that the jet stream this autumn was blowing much further north than normal.

Figure 6.24 The jet stream and British weather (*Source: The Independent*, 16 Nov. 1992)

14a What is a jet stream?
b How far do movements in the position of the jet stream explain the variations in weather described in Figure 6.24?

and south and broadly marks the boundary between polar air to the north and tropical air to the south (the reverse is true for the southern hemisphere). Figure 6.16 shows a sequence of ridges and troughs in the waves. The troughs (high-altitude, low-pressure areas) draw in cold air that descends in a clockwise direction to produce surface high pressure and dry, stable conditions. The ridges, in contrast, are areas of higher pressure and diverging air that lead to low pressure at the surface, unstable rising air, cloud cover and heavy precipitation.

In the tropics and sub-tropics, eastward-moving winds produce the Sub-Tropical Jet Stream about 20–30° north and south. This jet stream does not have such pronounced waves and has lower wind speeds. There is also an Easterly Equatorial Jet Stream that is more seasonal and influences the monsoon in the Indian sub-continent.

Jet streams and weather systems

Jet streams also guide depressions and storms from west to east in the middle latitudes. Occasionally, though, they migrate north and south. When a more southerly position is adopted by the Polar Front Jet Stream, it sends depressions into warmer southern waters that can lead to powerful storms, such as the one that had such devastating consequences in 1987 in England. On the other hand, when the Polar Front Jet Stream moves northwards, it can lead to a long period of fine summer weather, like the hot summer of 1976. This happens because high-pressure systems are trapped in the lower atmosphere, leading to a series of **blocking anticyclones**. These have a deflecting effect on low-pressure systems, much as whirlpools in a stream deflect currents of moving water. Meteorologists believe that the meandering of the jet stream across the Atlantic can also affect our weather for longer periods, for example, when the stream tended to follow a more southerly track in the 1980s (Fig. 6.24).

Changes in the position of the jet streams

Why the jet stream should meander is not very clear. Solar storms (short bursts of additional radiation from the sun) might affect the upper atmosphere and in turn influence the jet streams. Volcanic eruptions may also be important. The eruption of Mt Pinatubo in 1991, for example, led to huge quantities of gases and dust being ejected into the upper atmosphere that blocked the sun's energy and led to a cooling of the atmosphere (see Fig. 3.19). Human-made **global warming** could also exert an influence although, as we have seen in Chapter 4, the rate and degree of change is difficult to assess. The most popular idea relates to the influence of the oceans (see section 6.6). Meteorologists believe that the temperature of the sea influences the position of the jet streams and climate over several decades. For example, irregular 'pools' of warmer and colder water and the changing position of sea ice can alter over several years and then remain in a certain position for several more.

The 1987 storm in southern England

Thursday and Friday, 15–16 October 1987 are days people in the south of England will not forget for a long time. A storm occurred of such intensity that it was the worst in living memory and probably the most damaging since the 'Great Storm' of 1703.

On the evening of Thursday, a depression over the Bay of Biscay intensified and started to move north (Fig. 6.25). By midnight, powerful winds were already affecting the

Channel Islands and the area around the Isle of Wight. The storm gradually worked its way along the south coast during the night, resulting in incredible damage (Figs 6.26–6.27). Several people lost their lives largely as a result of buildings and trees that collapsed.

To the embarrassment of meteorologists, the storm was not well forecast. They thought that it would pass further south – over the north of France and so barely affect

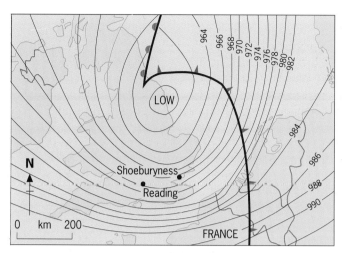

Figure 6.25 Weather map, 16 October 1987

Flying in the wake of the wind

WOODLAND beyond Windsor, Berkshire, looked as if a giant comb had been raked haphazardly through the trees. Patches of trees suffered a shell blast from the wind. On the unprotected edge of towns, the storm had left its imprint in a trail of flattened garages, torn fences and road signs bent backwards.

Caravan parks paid heavily for their uninterrupted sea view. One caravan park on Hayling Island, Hampshire, was totally wrecked and left to await the insurance assessor.

Shoreham airfield, Kent, was counting its damage. Two light aircraft lay smashed together, wheels in the air. The wings of two others had been torn by a trailer that the storm had turned into a missile. Altogether 27 aircraft were damaged, many beyond repair.

The trail of debris, torn off roofs and wrecked buildings continued along the south coast to Hastings, East Sussex…

southern England. This was a freak storm related to the jet stream tracking further south than usual. As a result, cold polar air moving south met a mass of warm tropical air moving north at the polar front. This then developed into a depression. On such occasions, accelerations of the jet stream draw up warm air faster than usual, which intensifies the low pressure and speeds up winds – as occurred in 1987.

— **?** —

15 State two ways in which the weather map in Figure 6.25 suggests that wind speeds were exceptionally high.

16 Study Figures 6.25–6.27 as well as the text.
a Attempt to classify the different impacts of the 1987 storm.
b Consider who benefited from the storm as well as those who suffered.
c Draw a matrix to show the effect of the storm on: • people's lives,
• the environment, • other areas.

17 Assume that you were given responsibility for co-ordinating the 'clean-up' operation after the 1987 storm. With reference to the text, Figures 6.26–6.27 and your own or others' experience, write a plan outlining your immediate and long-term objectives. Consider in particular the following areas:
a housing
b transport
c forestry
d energy supplies
e insurance claims.

DIY sales boom

CHAIN stores reported an unprecedented demand for tools and materials yesterday to cope with the trail of storm damage as householders prepared for the country's biggest ever DIY weekend.

Electricity cut to 3 million consumers

AN estimated three million consumers were affected by yesterday's power cuts and the Electricity Council has warned that for many it will be early next week before supplies can be restored.

Figure 6.26 How the press reported the impact of the 1987 storm

Figure 6.27 Storm damage

Figure 6.28 NOAA satellite image of the source of the Gulf Stream, January 1994. The computer image shows sea surface temperatures in the region around the Florida Peninsula, USA. The temperatures are colour-coded from grey and dark blue (0–2°C) through light blue, green, yellow and orange to red (22–26°C). At bottom-left is the Gulf of Mexico with large-scale eddy patterns visible.

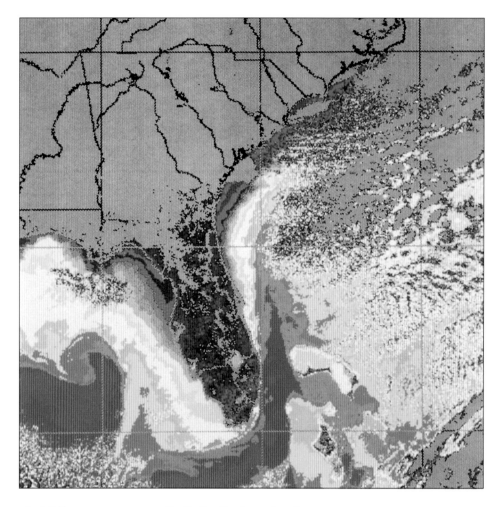

?

18 Study Figure 6.28.
a Identify the source area of the Gulf Stream.
b Describe the direction in which the Gulf Stream flows (see Fig. 2.27).

6.6 The oceans and global circulation

Many texts have overlooked or given little attention to the role of the oceans in influencing atmospheric movement and global climate. However, the oceans are important for two main reasons. Firstly, they provide water for the hydrological cycle and secondly, they absorb and redistribute energy (see section 2.5).

Beneath the surface of the oceans there is an equivalent 'marine atmosphere'. Like atmospheric winds, near surface currents occur which flow year after year such as the Gulf Stream (Fig 6.28). The oceanic equivalent to weather involves huge eddies (swirls of water up to 80 km across) that can move hundreds of kilometres and then die out.

Despite its latitude, Britain's mild climate exists largely because of the Gulf Stream which is warmed at its origins by the tropical sun. To the west of Europe, the current is called the *North Atlantic Drift*. This benefits the North Atlantic region as heat is released into air currents that move eastwards and produce a milder climate than would otherwise be the case over Europe.

Sea surface temperatures

About 80 per cent of solar energy is received by the oceans. Most of this is absorbed by the top 100 m of water, so clearly *sea surface temperatures* (*SSTs*) are of considerable importance in relation to global climate. One major area of research involves the Namias–Sabine hypothesis. This argues that irregular large-scale variations in sea surface temperatures, or **sea surface temperature anomalies** (**SSTAs**), strongly influence the atmosphere. These are believed to affect both local weather and more distant locations, although we are uncertain by how much. Scientists have

19 Use the pattern of temperature and rainfall maps of the British Isles in an atlas to explain how the Gulf Stream/North Atlantic Drift affects the climate here.

20 Essay: Assess the importance of the role played by the oceans in the world's climate.

21 Describe the pattern of the earth's wind systems. What part do they play in the movement of energy across the globe? Use diagrams and maps to illustrate your answer.

already seen the clear effects of SSTAs in the tropics, and in remote areas they follow responses called *teleconnections.* One such example is a link between SSTAs in the equatorial Pacific and pressure patterns over North America. The significance of the teleconnection varies with geographical location. Two key areas in equatorial waters are around the International Dateline and at 130°W. Researchers are using these climate models to integrate the sea surface anomalies and so produce better forecasting for months and seasons ahead.

The tropical part of the Pacific Ocean illustrates how unstable the ocean circulation can be. We have seen previously (see section 3.4) that occasionally the pressure systems, winds and ocean currents just to the south of the equator go into reverse. These **El Niño Southern Oscillations** have dramatic effects on the climate of western South America as well as much more widespread effects. On a much larger scale, change to the oceanic conveyor belt could have dramatic effects on the whole of the world. Some researchers have suggested that there is a link between the ice ages and the oceanic conveyor belt (see section 2.5). They consider that sudden climate changes may be associated with a breakdown of the conveyor belt, while global warming might result in more fresh water being released near the poles. This less dense, cold water would sink and therefore the North Atlantic would be colder. As a result, the climate of Britain and Europe could become more like Siberia rather than the Mediterranean. Understanding the fundamental links between atmospheric and oceanic processes could be vital if we are to predict the climate changes that may occur in the future as a result of global warming (see section 3.4).

Summary

- Atmospheric movement results from the uneven distribution of solar energy on the earth.
- There is a complex pattern of winds both in the upper atmosphere and near the surface. These are essentially the result of the tilt of the earth on its axis, the rotation of the earth and the uneven distribution of land and sea areas.
- The patterns of winds and the world's pressure belts determines the world's weather and climate.
- Winds move from high to low pressure areas. There is an inverse relationship between pressure gradient and wind speed. An increase in the pressure gradient results in stronger winds and vice versa.
- Winds occur at a number of scales, ranging from global geostrophic winds to local air movements affecting a few kilometres.
- Much has been learnt about global circulation in recent years particularly from satellites. This has led to new ideas and models. The basic model of atmospheric circulation today is based on research dating from the 1950s.
- Jet streams have an important influence on weather near the surface. Changes in the position and velocity of the jet streams can lead to intense storms.
- There are important seasonal variations in pressure and winds. The consequences for patterns of rainfall can exert a powerful influence on human activities like agriculture.
- There are important links between the oceans and atmospheric circulation. Large-scale variations in sea surface temperatures and circulation of water in the oceans could dramatically affect the world's climate.

7 Atmospheric moisture

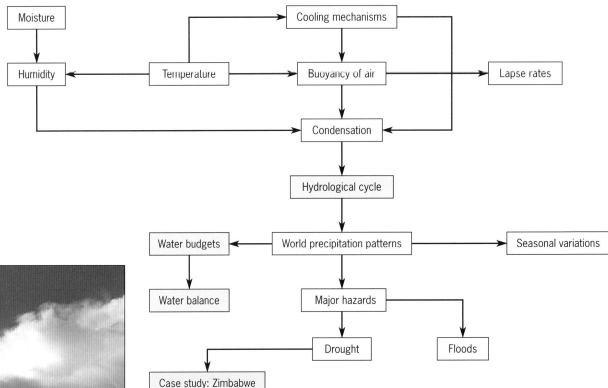

7.1 Introduction

The third aspect of the workings of the atmosphere, having examined energy and movement, is moisture. Water exists in the atmosphere in all three states, liquid, solid and gas, while changes between these states form a fundamental part of atmospheric processes. **Precipitation** collectively refers to water falling from the atmosphere in any form, such as snow, hail or rain.

Atmospheric moisture as clouds or precipitation gives a visible indication of atmospheric processes at work. Clouds signify **weather** that might occur in the near future that could influence human behaviour (Fig. 7.1). In fact, cloud patterns form an important part of any weather forecast (see section 1.4).

Moisture movements are also an integral part of energy flows in the atmosphere. We have previously seen that *water vapour* can carry **latent heat** from areas where there is a net radiant energy surplus near the equator to areas of net deficit at higher latitudes (see section 2.5).

Perhaps of the greatest significance is that atmospheric moisture is vital to life on earth. This is reflected in the distribution of the world's natural and climatic regions (Fig. 7.2). Although the world's natural regions reflect spatial variations in temperatures, each region also experiences different precipitation totals and regimes (annual distribution) (see Fig. 7.6). Consequently, spatial variation in the diversity of flora and fauna on earth is largely due to the uneven availability of water. Thus regular rainfall is more important than consistently high temperatures for producing the wealth of organisms in the world's most productive ecosystem, tropical rainforests.

Figure 7.1 Cumulonimbus clouds: the result of turbulent up-draughts of air often leads to storms, including heavy rain and hail

?

1   The study of cloud types is an important part of an airline pilot's training. What hazardous conditions might the clouds in Figure 7.1 suggest to a pilot?

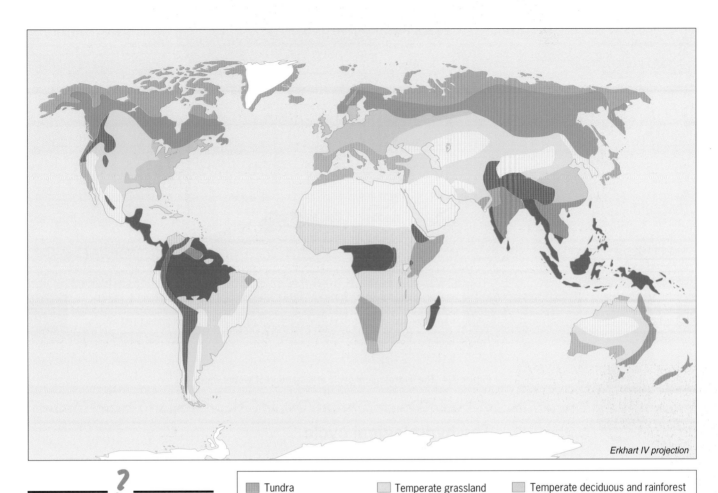

Erkhart IV projection

?

2a Compare Figure 7.2 with a world map of climatic regions in an atlas.
b Offer a general explanation for your observations.

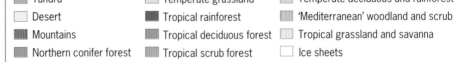

Tundra	Temperate grassland	Temperate deciduous and rainforest
Desert	Tropical rainforest	'Mediterranean' woodland and scrub
Mountains	Tropical deciduous forest	Tropical grassland and savanna
Northern conifer forest	Tropical scrub forest	Ice sheets

Figure 7.2 Major biomes of the world

Figure 7.3 Heat and humidity: suffering from an extreme of moisture in the atmosphere, The Gulf

We will also consider the many areas of the world where severe shortages of water have devastating consequences for people. Drought has led to loss of crops and millions have lost their lives in many African countries, and the famines in Sudan, Ethiopia and Zimbabwe in recent years have been in part due to the failure of rains. Clearly, it is important that we understand the atmospheric processes that lead to precipitation. Accurate prediction forms an important part of hazard management.

Extremes of moisture in the atmosphere can also lead to considerable discomfort for people (Fig. 7.3). We can do little to control the atmosphere and climate although there is one exception. We will consider, in Chapter 8, the extent to which artificial rainmaking is now possible. It is a technique that is based on a thorough understanding of atmospheric moisture and related processes.

7.2 Water movement in the earth–atmosphere system

The amount of water in the atmosphere at any moment is about 14×10^{12} tonnes (14 million million tonnes). However, this is only a tiny fraction of the total amount of water in the world. By a process known as the **hydrological cycle**, moisture in the atmosphere and on the earth's surface constantly moves from one part of the environment to another (Fig. 7.4).

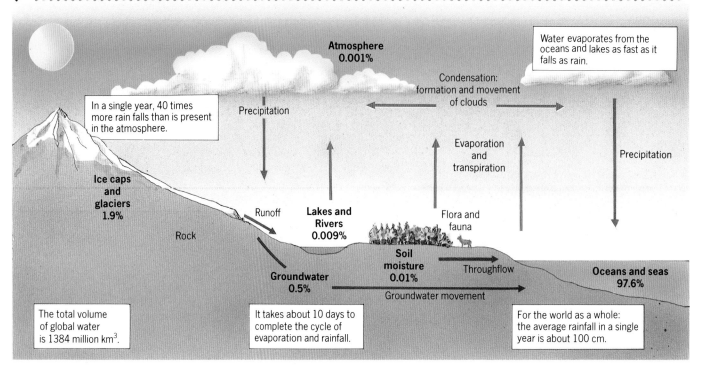

In a single year, 40 times more rain falls than is present in the atmosphere.

Water evaporates from the oceans and lakes as fast as it falls as rain.

The total volume of global water is 1384 million km³.

It takes about 10 days to complete the cycle of evaporation and rainfall.

For the world as a whole: the average rainfall in a single year is about 100 cm.

Figure 7.4 The hydrological cycle and location of global water

The hydrological cycle

A systems approach gives a clearer picture of the hydrological cycle and how atmospheric moisture interacts with other natural systems (Fig. 7.5). On a global scale, most **evaporation** takes place over the oceans, particularly in areas with high temperatures and where clear skies allow as much radiation as possible to reach the surface. Water vapour rises and is carried by winds, eventually leading to **condensation**, the release of latent heat (see section 2.5) and cloud formation. Thousands of kilometres may separate evaporation and precipitation and, while most clouds evaporate rather than produce precipitation, moisture eventually reaches the earth's surface to evaporate again and continue the cycle.

3 Explain why the hydrological cycle is an example of a *closed system* (see section 1.4 for *open system*).

4 How many *sub-systems* within the hydrological cycle are referred to in Figure 7.5?

5 Suggest what might be the smallest level of resolution (scale) at which the whole cycle could still operate in the natural environment.

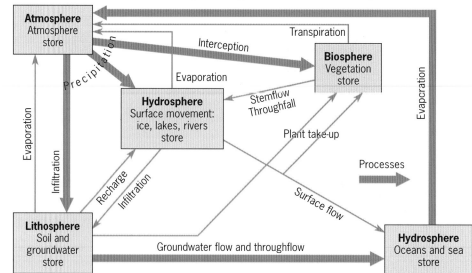

Figure 7.5 The hydrological cycle as a system

7.3 World precipitation patterns and water budgets

Water vapour is an invisible gas held in the air that accounts for the majority of atmospheric moisture. As warm air holds more moisture than cold air, we find most water vapour at lower latitudes. It then follows that just as there is an energy balance (see sections 2.4–2.5), so there must also be a **water balance** (balance between inputs and outputs of water). Like energy, moisture is also redistributed by surface flows and 'unseen rivers aloft' (Hilton, 1979). There are, however, considerable spatial and temporal variations in the distribution of precipitation at the global scale (Fig. 7.6). The general circulation of the atmosphere (see section 6.3) explains how moisture is carried from one location to another to produce this pattern.

The world pattern also indicates the importance of relief barriers and ocean currents. **Orographic** (linked to relief) influences are clear in areas like the south-west United States (see section 12.3). Here, the eastern side of the Rockies (**rain shadow**) is drier than the west. By the time eastward moving air currents reach states like Nevada and Arizona they have lost much of their moisture as rainfall in the mountains. The impact of oceans is seen where the cold Peruvian current off the west coast of South America is largely responsible for the aridity of the Atacama Desert. Warm, moist air moving to the east loses much of its moisture as it passes over the cold ocean and condenses to form mist and fog. By the time the air reaches land, it contains little moisture. Conversely, the North Atlantic Drift off the western side of Europe encourages higher precipitation. This is because the warm ocean current increases the ability of eastward-moving air to hold moisture.

?

6a Describe the pattern of world rainfall in Figure 7.6 referring to named areas.
b Select two or three areas and give reasons for their differences in rainfall.

7a Use Figure 6.8 and trace the outline of the major areas of high and low pressure. You could complete this exercise using similar maps in an atlas.
b Place your tracing over Figure 7.6. Describe and account for your observations.

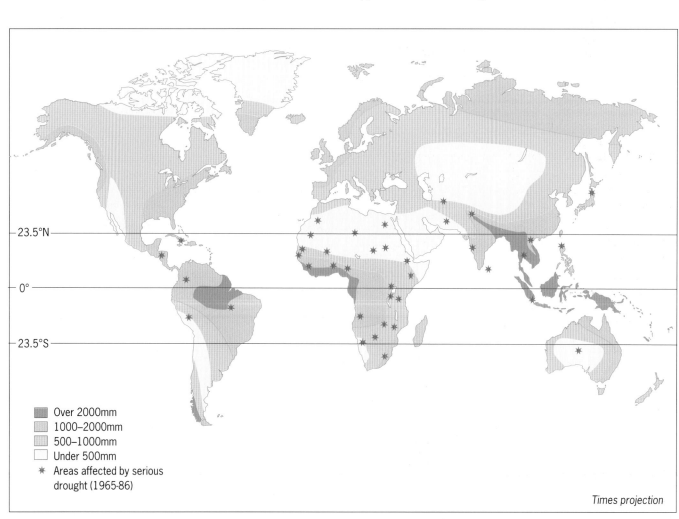

Times projection

Figure 7.6 World pattern of annual rainfall

- Over 2000mm
- 1000–2000mm
- 500–1000mm
- Under 500mm
- * Areas affected by serious drought (1965-86)

Water budgets

Global scale

As the hydrological cycle or system at the global scale is a closed system, there are no *inputs* or *outputs* of water and the budget is therefore zero. However, there is a spatial imbalance between areas that experience net evaporation (output) and areas that experience net precipitation (input) (Figures 7.7 and 7.8).

Figure 7.7 Latitudinal variations in evaporation and precipitation

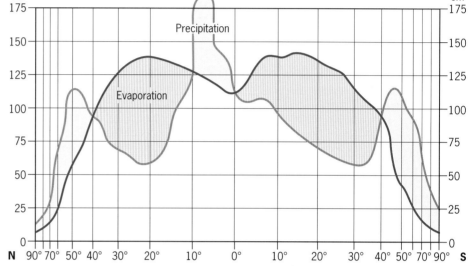

?

8 Study Figures 7.7 and 7.8. Refer back to Figures 6.8, 6.9 and 6.14 to account for the areas with the highest and lowest evaporation values.

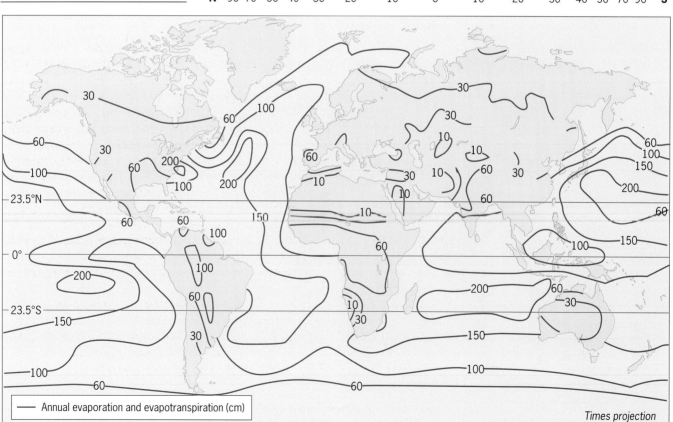

Figure 7.8 World pattern of evapotranspiration (*After:* Barry, 1969)

?

9a Using Table 7.1, describe the water budget of Australia.
b Suggest how this water budget may affect Australia's environment.

10a Draw a set of bar graphs based on the evaporation and precipitation data in Table 7.1.
b Place precipitation and evaporation results on the same bar for a given area. Use blue to represent precipitation and yellow for evaporation. Shade in blue the surplus of water, a deficit of water in yellow and the rest of the bar in green.
c Name the areas with the largest surplus and deficits of water. Try to account for your answers.
d Why does Asia have a smaller balance than Europe and North America even though they are in similar positions on the globe? (If you find this difficult, one reason might be suggested by section 12.4).

Continental scale

It is also possible to examine the **water budget** at the continental scale and for oceans (Table 7.1). However, these sub-systems are open and therefore have inputs and outputs. These are not in balance, so consequently water has to be redistributed by rivers and movement within the oceans. Clearly much precipitation falls over the oceans (a *store* in the system) to become available for evaporation again. In fact, considering their area, more precipitation falls here than over the continents.

The hydrological cycle also shows that a great deal of water vapour actually travels from the land areas towards the oceans (see Fig. 7.4). However, while the hydrological cycle over the oceans is relatively simple, the movement of water over land is more complex.

Table 7.1 The water budget of the continents and oceans

Area	Evaporation (millimetres)	Precipitation (millimetres)	Rivers or balance (millimetres)
Pacific Ocean	−1143	1219	+ 76
Australia	− 406	483	+ 77
Indian Ocean	−1372	1016	−356
Asia	− 381	610	+229
Europe	− 356	610	+254
Africa	− 508	660	+152
Atlantic Ocean	−1041	787	−254
Arctic Sea	− 127	254	+127
North America	− 406	660	+254
South America	− 864	1346	+482

Water balance

This is a similar idea to that of the water budget but relates to a smaller scale. This concept refers to the balance between the total inputs of precipitation (*P*) at a particular place and loss through outputs of **evapotranspiration.** Of all the precipitation received at one place, some is lost by evaporation while some is absorbed by plant roots and is then transpired through leaves. Evaporation and **transpiration** are very difficult to separate particularly when measurement is being attempted. These two processes are consequently combined and referred to as evapotranspiration. The remaining water is known as surplus water that can then infiltrate the soil or groundwater store or flow over the land as runoff towards the sea.

The rate of evaporation at any locality is dependent on a number of factors – principally temperature, length of day, hours of **insolation**, wind speed and **humidity**. When these factors combine to reach a point where maximum evapotranspiration occurs this is known as **potential evapotranspiration** (*PE*). This is an index that is very complex to calculate but it simply relates the amount of precipitation required for maximum evaporation to take place (assuming vegetation is present) to the actual water available.

When *actual evapotranspiration* is less than potential evapotranspiration then a place is said to experience a *moisture deficit* (*MD*). An area experiences a *moisture surplus* (*MS*) when the opposite occurs. Surplus moisture can replenish the soil and make up for a moisture deficit while any remaining moisture drains away.

The interaction between these variables can be summarised by a simple equation:

$$Water\ balance = P - PE + (MS - MD)\ where:$$
P = precipitation $\quad$ *MS* = moisture surplus
PE = potential evapotranspiration $\quad$ *MD* = moisture deficit.

Water balance

The pattern of water availability at any place over a year is often shown graphically. The *water balance diagram* in Figure 7.9 shows that from March to late September *PE* exceeds *P*. As a result there is initially a period when the moisture already in the soil is utilised by vegetation. As outputs of moisture are greater than inputs during the summer months a moisture deficit eventually exists. From October to early January *P* is greater than *PE* so the soil is recharged to the extent that by early January there is a moisture surplus that lasts to mid-March.

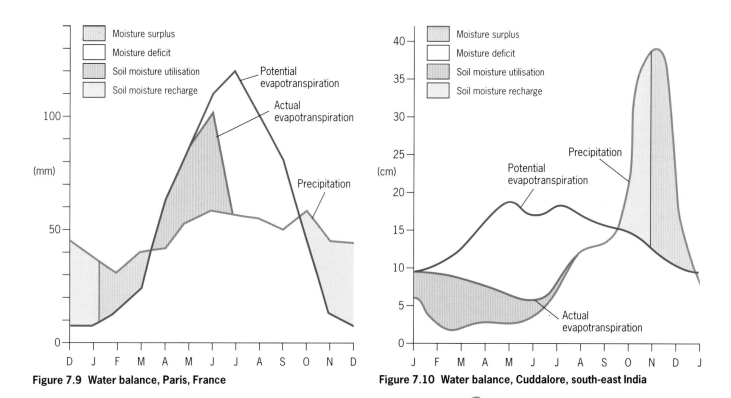

Figure 7.9 Water balance, Paris, France

Figure 7.10 Water balance, Cuddalore, south-east India

?

11 Study Figure 7.10.
a Explain why there is a moisture deficit between January and late September.
b Describe and explain the period from late September to late December.

12 How could farmers in this part of south-east India benefit from such a graph?

7.4 Drought

A drought is simply a continuous period of dry weather. In many parts of the world, drought is a serious problem which affects a range of human activities and can also lead to the hazard of fire (see Fig. 7.6).

In the British Isles, a distinction is made between absolute drought, referring to a period of at least 15 consecutive days each with less than 0.25 mm of rainfall, and partial drought. The latter is a period of 29 consecutive days during which the daily average does not exceed 0.25mm, although some days may have slight rain. Such a precise definition would be unsuitable in a very arid part of the world and consequently definitions of drought vary between countries. In Africa, a different approach is more appropriate. Drought is defined as a period when evapotranspiration exceeds precipitation and soil moisture is depleted to the extent that crops and much natural vegetation cannot grow.

Drought in Zimbabwe 1989–92

Zimbabwe has been described as the bread basket of southern Africa. It is usually a country rich in wildlife and able to produce food crops to support its population with a food surplus. The country has two of Africa's major rivers, the Limpopo to the south, and at its northern border the Zambezi with one of Africa's best known natural features: the Victoria Falls (Fig. 7.11). In the late 1980s the country was stricken by severe drought. This illustrates well the problems that can be created when there is a considerable imbalance between inputs of rainfall and outputs from evapotranspiration and runoff.

Impacts of the drought

The wet season is usually from October to April but during 1989–92 the rains did not start until December and were substantially below normal levels, causing severe drought. Poor rains affected almost the whole country in the 1991–2 season with a number of devastating consequences (Figs 7.12– 7.14).

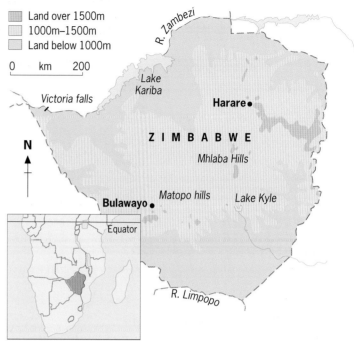

Figure 7.11 Zimbabwe: relief

Legend:
- Land over 1500m
- 1000m–1500m
- Land below 1000m

0 km 200

A country dying of thirst

IN THE HIGH veld rivers are reduced to trickles, reservoirs to puddles. In the low veld they are just dry. We went for miles through a ghostly, lifeless, leafless landscape of ash-grey skeleton scrub.

The Limpopo when full is at a guess twice the width of the Thames at Westminster. It's not full. A few days ago I stepped across the Limpopo without getting my shoes wet. Cracks have broken the river bed into a crazy paving of pieces about 45cm across.

The Victoria Falls are still on tap but Kariba is at its lowest level since it was commissioned in 1959. It is expected soon to reach a level at which electricity production will have to cease. This would mean increased dependency on the country's four thermal power stations. Would these be able to generate enough to keep the country going?

The dam at Lake Kyle is 350m long and 60m high. It was built in 1960 and the reservoir took 15 years to fill with 1400 million m³. Then for seven years it overflowed. It stopped overflowing in 1982 and now Lake Kyle hardly exists. And this is the great reservoir which is supposed to irrigate the agricultural and economic riches of the Hippo valley citrus estates and the Triangle sugar estates.

Maize is the country's staple food crop; this year the crop has failed completely in all eight provinces. By the end of the year half the national cattle herd will have starved or been slaughtered.

In Bulawayo tap water supplies are restricted to a few hours a day, and discs have been inserted into the pipes to restrict the flow even during those hours. Water was first rationed to 600 litres a day per household, which was cut to 400 and now by a further 10 per cent to 360. In Mutare its down to 100 litres per household.

Figure 7.12 How the press reported the 1989–92 drought, Zimbabwe (*Source: The Guardian*, 1993)

Zimbabwe's GDP fell by eight per cent in 1992. Agricultural sector output fell 40 per cent generally, and that of maize and sugar alone by 75 per cent. Agro-based industries – a significant proportion of the economy – suffered accordingly. Industry suffered because of electricity rationing, mainly due to Kariba's low water level. Water rationing was introduced in some areas. There was speculation about the evacuation of Bulawayo, Zimbabwe's second city, because of the chronic shortage of water in the area.

Hit worst of all were villages in communal areas. Without further food stocks and facing the sight of their cattle dying, many families were forced to move in with relatives in cities or on commercial farms.

Figure 7.13 Effects of the 1989–92 drought in Zimbabwe (*Source: British Overseas Development, May 1993)

Figure 7.14 Zimbabwe's cattle die from lack of water

Drought in Zimbabwe

Table 7.2 Total annual rainfall for Harare, Zimbabwe 1960–94 (*Source:* University of East Anglia, Bureau of Meterorology, Harare)

Year	Total rainfall (mm)	Year	Total rainfall (mm)
1960–1	654	1977–8	1082
1961–2	955	1978–9	1088
1962–3	822	1979–80	832
1963–4	750	1980–1	800
1964–5	569	1981–2	963
1965–6	745	1982–3	645
1966–7	717	1983–4	548
1967–8	676	1984–5	674
1968–9	604	1985–6	1064
1969–70	1123	1986–7	901
1970–1	679	1987–8	580
1971–2	615	1988–9	794
1972–3	871	1989–90	646
1973–4	893	1990–1	674
1974–5	1272	1991–2	592
1975–6	736	1992–3	906
1976–7	818	1993–4	763

POSITIVE ACTION

There was a spirit of co-operation and determination throughout Zimbabwe, and unprecedented co-operation throughout the region. South African and Mozambican ports, railways and roads became lifelines to the interior and performed magnificently.

A system designed to buy grain locally and to export it – spearheaded by the Grain Marketing Board – went into reverse most efficiently. Over two million tonnes of maize were imported in the twelve months from May 1992, 80 per cent moved by rail. Distribution to the local level sometimes ran into problems but by and large the food got to where it was needed.

A national water strategy is now also under review. Planning for power lines to Botswana to link up with the South African grid are under way, giving Zimbabwe greater options in future.

NGOs [Non-Governmental Organisations] liaised closely with government ministries. The department of health, for example, was closely involved in setting up child feeding pro- grammes and the department of social welfare helped NGOs organise drought relief distribution.

Figure 7.15 Zimbabwe copes with the drought (*Source:* British Overseas Development, May 1993)

Lessons learnt from the drought

Despite the economic and social costs of the drought, Zimbabwe coped exceptionally well and people learnt many valuable lessons (Fig. 7.15). Although there was a great deal of international aid, the Zimbabwean authorities established emergency drought relief measures which should reduce the impact of drought in the future. This included a supplementary feeding programme for about a million needy children set up by the Department of Health.

The rains finally returned at the end of 1992. They were patchy at first but by December most regions had adequate rainfall for agriculture to recover. 1993 was a wetter year and farmers reaped better harvests. It will take several wet seasons, though, for reservoirs to be replenished and for the cattle herds to build up again. If another drought were to occur there is no doubt that Zimbabwe and the region would be much better prepared and in a stronger position to manage a shortage of water and the upheaval it causes.

13a Plot a line graph based on the data in Table 7.2. Draw a line representing the mean annual rainfall for the period 1960–94
b What was exceptional about the period 1989–92?

14 Use Figures 7.12–7.14 to identify and classify the different impacts of the 1992 drought in Zimbabwe.

15 To what extent do you think the experience of individuals, organisations and the government gained during the drought will help with the management of water shortages in the future?

16 Imagine that you had been in Zimbabwe during the 1989–92 drought. Write a report outlining what you see to be the main short and long-term priorities to overcome the effects of the drought and reduce the impact of drought in the future. Use Figure 7.15 and the text to help you.

17 Essay: With reference to a specific country or part of the world (other than Zimbabwe), state what you understand by the term drought, and outline the effects a drought can have on human activities.

7.5 Humidity and condensation

Defining humidity

Humidity refers to the concentration of water vapour in the atmosphere. Two factors control the amount of water vapour present at any moment: temperature and the availability of water. The mass of water vapour per unit volume of air is known as *absolute humidity*; it is expressed in g/m^3. *Saturated air* refers to the maximum amount of vapour at a given temperature being held in the air (Fig. 7.16).

Another way meteorologists look at humidity is to consider the pressure of water vapour amongst other gases in the atmosphere. This is known as **vapour pressure** and in saturated air, as *saturation vapour pressure*. Vapour pressure ranges from about 2mb over Siberia in the winter to about 30mb in the tropics in the summer.

18 Study Figure 7.16.
a Describe the relationship between air temperature and absolute humidity.
b State the two ways in which the unsaturated pocket of air at X could be brought to saturation.

19 State three meteorological elements that control the rate of evaporation.

20 If a saturated pocket of air at 15°C contains 13g of water, calculate the relative humidity if the air temperature was raised to 26°C.

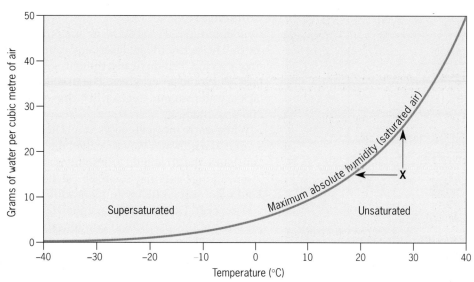

Figure 7.16 The relationship between air temperature and absolute humidity

The term **relative humidity** refers to the total amount of water vapour present in a mass of air, expressed as a percenatge of the total amount that would be present if the air was saturated at that temperature (the ratio between vapour pressure and saturation vapour pressure). It should follow that this varies with absolute humidity and with temperature.

ABSOLUTE HUMIDITY
Saturated air (100%) at 20°C contains 17g of water vapour per cm³.
If the air only contained 10g at 20°C:
$$\text{Relative humidity} = \frac{10}{17} \times 100 = 59\%$$

TEMPERATURE
Saturated air (100%) at 10°C contains about 8g of water vapour.
Saturated air (100%) at 20°C contains about 17g of water vapour.
If the air temperature was raised from 10°C to 20°C with no addition of water, the relative humidity would fall to 47%:
$$\text{Relative humidity} = \frac{8}{17} \times 100 = 47\%$$

Humidity and people
In the British Isles we rarely experience extremes of humidity for very long. However, in other parts of the world very humid conditions or extreme aridity can make life difficult for the people who live there (Fig. 7.17, see Fig. 7.3). Visitors and tourists who are not used to such conditions are particularly vulnerable, as it is easy for people on holiday to be active and to push themselves beyond safe limits in the sun. This can result in heat exhaustion, which is caused by a fall in the body's fluid level, and salt depletion caused when we lose salt through perspiration. In severe cases, this can lead to heatstroke which can kill.

21 How can continual humid and hot conditions hinder development in some countries? (See Fig. 7.3.)

22 Write a section for an information book for a hot country to advise tourists how to avoid heat exhaustion and heatstroke.

It was still winter. By Gulf standards the weather was pleasantly cool. Yet by nine in the morning one could feel the damp heat building in the air for a tropical storm that never came. By ten, I was soaked in sweat. My room was littered with shirts that had gone stiff with salt. I cannot imagine how the migrant construction workers who swarmed like locusts round the skylines of the city were able to keep going in that suffocating air. In this climate the natural thing was to live a life of unpunctuality, idle conversation and frequent refreshment with the coffee ceremony; but now the Gulf was determinedly flying in the face of nature.

Figure 7.17 Humid conditions in the Gulf of Aden (*Source*: Rabin J, *Arabia through the looking glass*, 1979, William Collins)

Figure 7.18 Hang-glider rising on a thermal

Condensation

Condensation is a vital atmospheric process which is of great significance to life on earth. The lack of life in desert environments reflects the lack of condensation and precipitation. Paradoxically, though, some of the largest amounts of water vapour pass over some desert areas. The problem of aridity is therefore not always due to a lack of water vapour, but lack of a change of state to liquid through condensation and precipitation.

When warm moist air cools, it reaches a temperature where the air becomes saturated with water vapour, called the **dew** or **condensation point**. If cooling continues, the air should lose some water vapour as it condenses and reverts to water. Strange as it may seem, air can be *supersaturated*, or in other words it can hold water vapour below condensation point and have a relative humidity of over 100 per cent. Experiments have shown that effective condensation relies on the presence of tiny particles or **condensation nuclei** around which water can condense. They include natural particles like dust from volcanic eruptions and salt over the sea. The latter attracts water very easily, encouraging condensation even before saturation; they are known as **hygroscopic particles**. Industrial pollutants can also encourage condensation with up to several million smoke particles occurring per cubic centimetre.

7.6 Cooling mechanisms and the buoyancy of air

Cooling can occur either by horizontal movement, called **advection**, or by vertical convective movements. Advective cooling generally takes place over a cold land or sea surface and encourages mist and fog to develop. In contrast, vertical movement is generally more important for encouraging condensation and cloud formation (see Fig. 7.1). Air may rise due to its natural buoyancy associated with its warmth and humidity, or it can be forced to rise by relief or to pass over a cold and denser air mass.

If you have ever climbed in a mountain area you would have noticed a fall in temperature. This change with height in the **troposphere** is known as the *environmental* **lapse rate** (*ELR*) which varies spatially and temporally. Within this general environmental air, pockets of air rise and fall. Rising warm air pockets or **thermals** are used by glider pilots to gain altitude (Fig. 7.18). As a thermal gains height it cools at the *adiabatic lapse rate* (*ALR*). At higher altitudes air pressure is lower and the rising air will tend to expand. Energy is then lost when it is converted, as a result of expansion, into kinetic energy. Additional cooling may also occur as a result of conduction with surrounding environmental air. If this rising air is not saturated, it cools at a virtually fixed rate in the lower atmosphere of about 10°C for every 1000 m. This is known as the *dry adiabatic lapse rate* (*DALR*).

Rising air may eventually reach condensation level and become saturated. Once condensation occurs, latent heat is released and the energy that originally

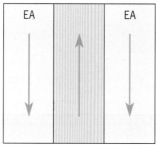

Column has a tendency to rise, being warmer than surrounding air

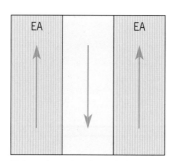

Column has a tendency to be stable or sink, being colder than surrounding air

Warm air

Cold air

EA = Environmental air

Figure 7.19 The relationship between air temperature and buoyancy

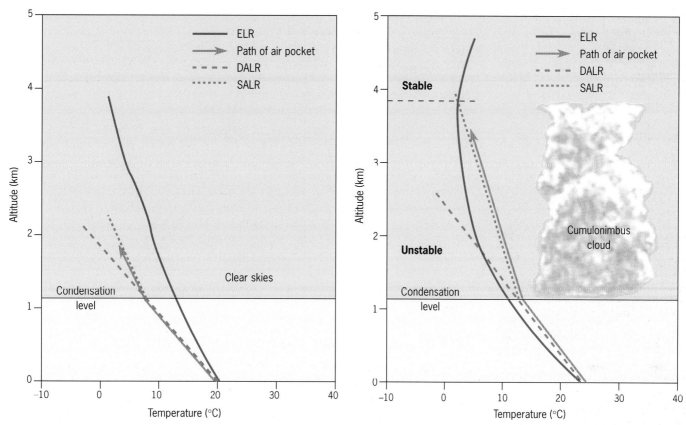

Figure 7.20 Lapse rate curves showing atmospheric stability

Figure 7.21 Lapse rate curves showing atmospheric instability

23 Study Figure 7.22. Explain briefly what is happening to the air in each of the three layers A, B and C.

24 What important meteorological changes occur at X and Y in Figure 7.22?

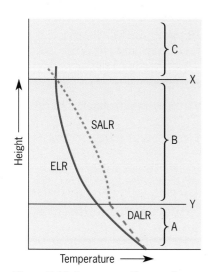

Figure 7.22 Lapse rate diagram for test

produced the water vapour is now released as the vapour turns back into water. This warmth added to the pocket of air partly compensates for the cooling with height. The fall in temperature is consequently less than for dry air. The rate at which this air cools is known as the *saturated adiabatic lapse rate* (*SALR*). This is about half the DALR near the ground (6°C per 1 000 m) but nearly equal near the top of the troposphere (at very high altitudes air is much drier so the release of latent heat is reduced and the rate of cooling is quicker).

The buoyancy or **stability** of air depends on its density in relation to the surrounding air. This is inversely related to temperature. In other words, air warmer than its surrounds tends to rise and be unstable, while air colder than its surrounds tends to sink or remain where it is and be very stable (Fig. 7.19).

Under conditions of atmospheric stability, a pocket of air will not tend to rise. Figure 7.20 shows that the environmental air is warmer than a pocket of air rising at the DALR and the SALR, although there is a path curve. The pocket is therefore not buoyant. With atmospheric **instability**, the reverse is now the case and the pocket of air is very buoyant (Fig. 7.21). Towering clouds then tend to develop above condensation level. You will notice that above 3800 m the pocket becomes stable again as the SALR becomes less than the ELR. *Neutral stability* (Fig. 7.23) is a situation where the pocket of air has the same temperature as the surrounding environmental air; under such conditions air will tend to remain where it is as long as buoyancy is the only control.

Bodies of air are often forced to rise over hills or mountains. Figure 7.24 shows air that is stable near the ground but becomes unstable at an altitude of about one kilometre; the air mass is said to be *potentially unstable*. In this case, the lapse rate of the air is initially less than the ELR but at one kilometre the DALR becomes greater than the ELR. When instability occurs, it is conditional on the air becoming unstable and saturated as a result of vertical uplift; this is therefore referred to as *conditional instability*.

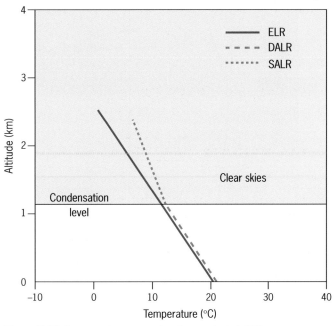

Figure 7.23 Lapse rate curves showing neutral stability

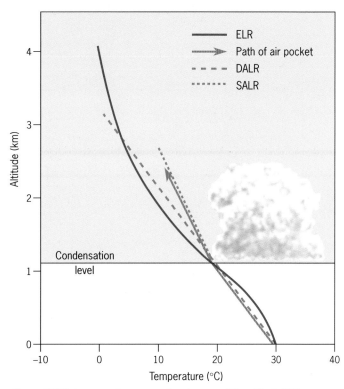

Figure 7.24 Lapse rate curves showing conditional instability

25 Why does the cloud in Figure 7.24 not extend above 3800 metres?

26 Use your own words to outline the difference between neutral, potential and conditional instability.

27a Draw a graph as follows to illustrate conditional instability:
• vertical axis from 0 to 11 km altitude
• horizontal axis from –60°C to +40°C
• draw a straight line for the ELR assuming a surface temperature of 20°C and a rate of fall in temperature with height of 8°C per kilometre
• draw your own path curve for the DALR and the SALR and state the altitude at which the air becomes unstable.
b Describe the weather this might lead to in the summer over an upland area like the Pennines.

28 Essay: Outline the causes and effects of atmospheric stability and instability.

Summary

• Water exists in the atmosphere in three states: liquid, solid and gas. Change between these states is fundamental to many atmospheric processes.

• Atmospheric moisture is vital to life on earth, producing spatial variations in flora and fauna and influencing human behaviour.

• The hydrological cycle as a system provides a useful framework for studying atmospheric moisture. This system interacts with other global systems such as ecosystems.

• Spatial variations in precipitation at the global scale are related to the general circulation of the atmosphere. Relief and ocean currents are also important influences.

• The concept of the water budget can be used to show spatial variations in the balance between precipitation and evaporation.

• A drought is a continuous period of dry weather leading to crop failure and damage to natural vegetation cover. Severe droughts can lead to loss of life and be a serious economic drain on a country.

• Humidity is controlled by two factors: temperature and the availability of water. Warm air can hold more moisture than cold air.

• Condensation is related to temperature and humidity and in the natural environment is dependent on the availability of hygroscopic particles.

• The buoyancy or stability of air is dependent on its temperature and density in relation to the surrounding air. Air can become unstable when it is heated or forced to rise over an upland area or another air mass.

8 Precipitation

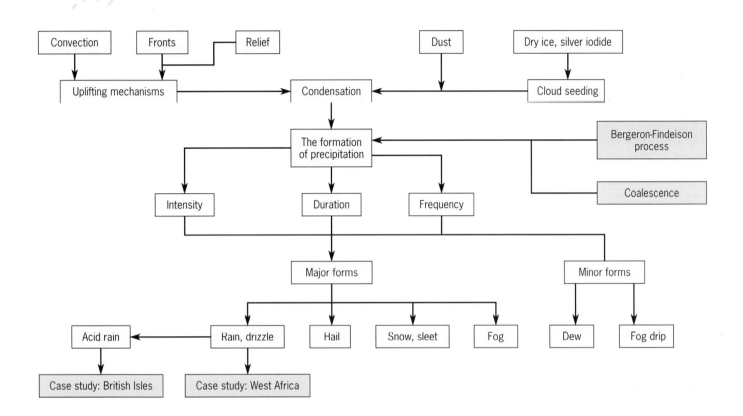

Convection → Uplifting mechanisms
Fronts → Uplifting mechanisms
Fronts, Relief → Uplifting mechanisms
Uplifting mechanisms → Condensation
Dust → Condensation
Dry ice, silver iodide → Cloud seeding → Condensation
Condensation → The formation of precipitation
Bergeron-Findeison process → The formation of precipitation
Coalescence → The formation of precipitation
The formation of precipitation → Intensity, Duration, Frequency
Intensity, Duration → Major forms
Frequency → Minor forms
Major forms → Rain, drizzle; Hail; Snow, sleet; Fog
Minor forms → Dew; Fog drip
Rain, drizzle → Acid rain
Acid rain → Case study: British Isles
Rain, drizzle → Case study: West Africa

Figure 8.1 Intense rainfall, Virac, Philippines

8.1 Introduction

Precipitation refers to the many ways in which moisture falls from the atmosphere. Some parts of the world, like north-east India, have an excess of rainfall leading to frequent floods . If we lived in parts of the tropics like Indonesia or the Philippines, we would experience rain virtually every day of the year! (Fig. 8.1) Details of precipitation form an important part of **weather** forecasts. However, precipitation is not always predictable and sometimes arrives suddenly and at great intensity creating hazardous conditions. Blizzards often create chaos to transport in the winter. We will consider the extent to which precipitation influences human activities, like transport, farming and tourism.

We will examine the different forms of precipitation like rain, snow, hail and sleet and consider how they develop. It may seem that there is little we can do to control the amount of rain falling in an area or to reduce a hazard like a hailstorm. However, an understanding of the formation of precipitation has led to a degree of control over these atmospheric processes. Conversely, human activities have upset the temperature and composition of the atmosphere so that, for example, **acid rain** reflects a disturbance to moisture in the earth–atmosphere system.

8.2 The formation of precipitation

The study of clouds is largely beyond the scope of this book; they are very complex and there are many different types. Clouds are composed of water droplets or ice crystals or a mixture of both, depending largely on temperature and air currents within the clouds. A high proportion of water droplets leads to clouds with clearly

Figure 8.2 Cirrus clouds

defined edges like *cumulus* (see Fig. 7.1); these tend to be low or medium altitude clouds. A predominance of ice crystals gives a fibrous, wispy appearance as with high altitude *cirrus* clouds (Fig. 8.2).

Many people hold a common misconception that **condensation** leads to precipitation. This is not the case, though, with all clouds. Water droplets in clouds are often very small, starting at 0.1 mm in size and growing to a maximum of about 5 mm. These droplets are held up by air currents and movements. A large cloud may hold thousands of litres of water with a huge amount of potential energy.

Clearly, droplets will fall to the ground when the pull of gravity exceeds the ability of air currents to keep them held in the atmosphere. The critical issue is how tiny droplets reach such a size that causes them to fall. We can conclude that the droplets act as **condensation nuclei** and encourage further growth. However, research shows that clouds form and rain falls more quickly than this process would allow.

The Bergeron–Findeison process

In the 1930s two Norwegian scientists, Bergeron and Findeison, developed one of the main theories of precipitation (Fig. 8.3). It assumes that ice crystals and water droplets coexist in clouds. This has been confirmed by radar observations and more recently by satellites. Condensation occurs relatively easily in comparison with freezing in clouds, as there are more condensation nuclei than freezing nuclei. Spontaneous freezing occurs when small water droplets are super-cooled to –40°C. The freezing nuclei that are needed to encourage ice crystals to form mostly consist of fine soil particles and dust from volcanic eruptions. Following formation, large ice crystals can splinter in turbulent air currents in clouds to form small splinters that can act as additional nuclei.

Moving ice crystals in large clouds, such as *cumulonimbus*, are believed to grow by further condensation and freezing. *Water vapour* can condense very rapidly and

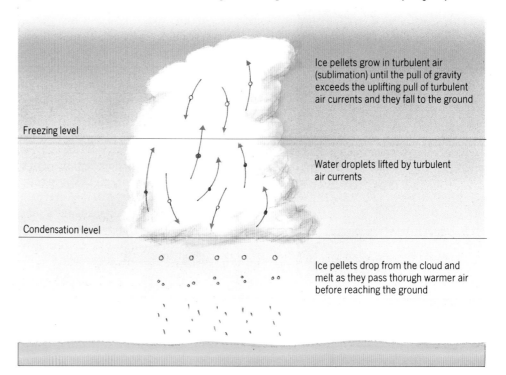

Figure 8.3 Bergeron-Findeison mechanism of precipitation formation

Figure 8.4 Example of cumulus clouds required for seeding: before seeding the clouds need to have their base at approx. 3300 m, with their top at 5500m

Figure 8.5 After dry ice is dropped into a cloud, rain drops develop and heavy rain will fall

1 Look at Figures 8.4 and 8.5. What evidence is there that cloud seeding might have taken place?

turn straight from a gas to ice. This process is often called **sublimation**, although strictly this refers to direct **evaporation** from ice. This process is responsible for much of our rainfall in the British Isles. In temperate latitudes, large quantities of precipitation only fall from clouds that extend above freezing level; lower clouds tend to produce drizzle.

Cloud seeding

Cloud seeding is the artificial process of rain making and is largely based on the Bergeron–Findeison theory. This is because the basis of the activity lies in the importance of freezing nuclei. Supercooled clouds with water between –5°C and –15°C are *seeded* from aircraft with solid carbon dioxide pellets, known as dry ice, and silver iodide. **Latent heat** (see section 2.5) is released, which in turn causes clouds to expand and absorb more water vapour from the surrounding air. Condensation can increase the clouds' water content by as much as 25 per cent, which can then fall as rain.

The cloud seeding technique was mostly developed in the USA and Australia, with much of the early work taking place just after World War 2. There were successes, but equally there were occasions when seeding led to a reduction in rainfall. The latter happens when seeding occurs too early and prevents the updrafts of air that would encourage cloud development. Successful seeding therefore depends on the amount of seeding materials used and selecting the appropriate meteorological conditions and clouds. Research, however, continues, particularly in parts of the world where very dry climates prevail.

People can use cloud seeding as part of their management program to cope with drought or very arid conditions. This happened during the 1980s and 1990s in Israel, when scientists had positive results. Weather radar at the Israel Rain Enhancement Project's headquarters in Tel Aviv was used to detect clouds suitable for seeding, that is, average rain clouds over Israel which contain about 500 000 m³ of water. Subsequently, cloud seeding increased the height of some clouds by more than a kilometre and so doubled water content and rainfall yield. These results suggest that rainfall totals in some areas may be increased to provide vital water for irrigation projects.

The use of radar and satellite technology

The early 1990s were hot and exceptionally dry years for northern Thailand, even during the wet season. Rains were light and late. As a consequence, the water level in reservoirs lowered, threatening irrigation and rice production and HEP schemes. Like Israel, many Asian countries have used radar to find suitable clouds and plan cloud seeding missions; in Thailand up to fifteen a day. There is a problem with radar, though, because it plots clouds *with* rain droplets, although cloud seeding is most effective before the droplets form. A Japanese weather satellite has provided a solution of an infra-red sensor which shows clouds at different stages of development including embryonic clouds. This indicates suitable clouds for seeding and scientists can select the most appropriate chemicals. This satellite also extends the planning time for missions from 2 to 3 hours to a day. Indonesia is now interested in this research and, if success continues, other south-east Asian countries are likely to follow suit. Seeding does not solve drought but it can reduce the problem.

Scientists have also experimented with cloud seeding to suppress lightning and to influence the development and direction of **hurricanes**. Even though it is too early to draw conclusions, it will never really be possible to control large scale atmospheric systems like hurricanes.

The coalescence theory

In some parts of the world, particularly in the tropics, heavy rain falls from clouds which have few ice crystals and cloud-top temperatures up to 5°C. This means that other mechanisms for the formation of precipitation must exist. To explain this, two meteorologists, Simpson and Mason, developed the *coalescence theory* (Fig. 8.6). In the tropics strong **thermals** help the development of precipitation, as large clouds can contain several turbulent currents. Eventually, droplets reach a size where gravity pulls them to the ground.

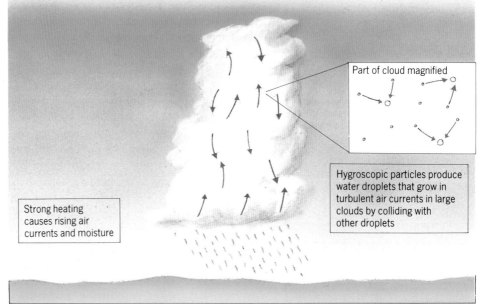

Part of cloud magnified

Strong heating causes rising air currents and moisture

Hygroscopic particles produce water droplets that grow in turbulent air currents in large clouds by colliding with other droplets

2 Use Figure 8.6 to explain in your own words why heavy rain falls from clouds with few ice crystals.

3 How does atmospheric instability encourage the formation of large rainclouds?

Figure 8.6 The coalescence mechanism of precipitation formation. Developed by Longmuir.

It is quite possible that both the processes described in this section could be at work at the same time. The Bergeron–Findeison would take place in the upper parts of clouds, while the coalescence process occurs at lower altitudes. In recent years however, scientists have discovered a problem with these mechanisms. Sometimes substantial rain falls from small clouds at low altitudes and below freezing point. Recent research indicates that higher altitude clouds may be seeding the lower altitude clouds with ice crystals. These act as condensation nuclei and so encourage precipitation development.

8.3 Forms of precipitation

In section 8.2 we studied an outline of how precipitation develops and a background to precipitation forms. This section will consider different types of precipitation, how they develop, their properties and how they influence human activities, particularly in extreme conditions.

Rain

Rain is classified as droplets 0.5–2 mm in size. The *intensity* of rainfall refers to the amount of rain occurring in a period of time. *Duration* refers to the length of time over which rain falls. There are three types of rainfall, reflecting the different atmospheric conditions under which they develop. In each case a mechanism encourages moist air to rise. As the air reaches **condensation level**, clouds develop, leading to rainfall. It is important to appreciate that these mechanisms are not independent of each other. For example, rainfall could be enhanced as a **depression** passes over an upland area combining frontal and **orographic** influences.

Convectional rainfall

Convectional rainfall results from strong upward-moving, buoyant air currents (Fig. 8.7). In the British Isles, convectional rainfall is particularly common in the summer when there are consistently high temperatures, resulting in short-lived showers and thunderstorms. These most notably occur in the south-west of England where such storms have developed at sea and tracked eastwards towards the land. Very intense convectional storms are common in the Alps during the summer. Often, they trigger other events such as landslides, when material on a mountainside is lubricated and has additional weight after a storm (Fig. 8.8).

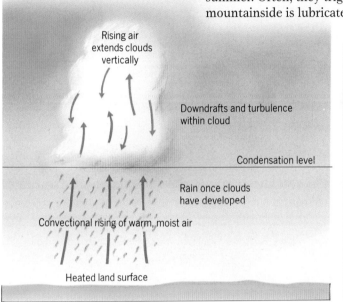

Figure 8.7 The formation of convectional precipitation

Figure 8.8 Mudslide in the Alps destroys buildings following intense convection rainfall, Tartano, Italy

Convectional rainfall is also very common in the tropics. Here the same convectional processes are added to by convergence of air currents at the **Inter-Tropical Discontinuity (ITD)** (see section 6.3). This gives rise to towering cumulo nimbus clouds that can be as wide as 10 km. Larger cloud masses may extend as much as hundreds of kilometres wide and can produce intense rainfall over thousands of square kilometres (see Fig. 6.14). There is a more detailed examination of tropical weather systems and precipitation in Chapter 10.

Orographic rainfall

Orographic rainfall is associated with upland areas that encourage warm, moist air to rise. This leads to conditional instability. We can find a particularly striking example of this in Tenerife in the Canary Islands (Fig. 8.9). The Trade winds that approach from the north over the Atlantic bring about 450 mm of rainfall each year, most of which falls on the north side of the island (Fig. 8.10). When the air currents reach a height of 1500–1800 m, condensation leads to a layer of cloud (Fig. 8.11). By the time the air masses have crossed the uplands, most moisture has already been released. The leeward side of mountainous areas are consequently relatively dry and are referred to as the **rain shadow**. The south side of Tenerife is in the rain shadow and consequently has a semi-arid landscape (Fig. 8.12). In contrast, the north has a denser vegetation cover and most of the island's main banana plantations (Fig. 8.10).

Of all the precipitation that falls on the island, about 72 per cent is lost through

Figure 8.9 Tenerife, Canary Islands (Spain)

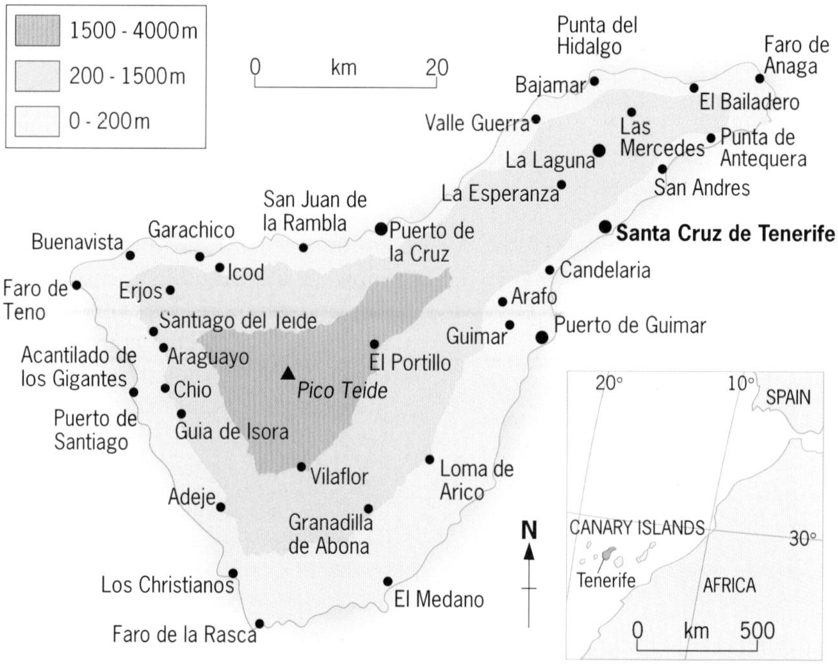

Figure 8.10 Tenerife, north side: clouds rising over denser vegetation

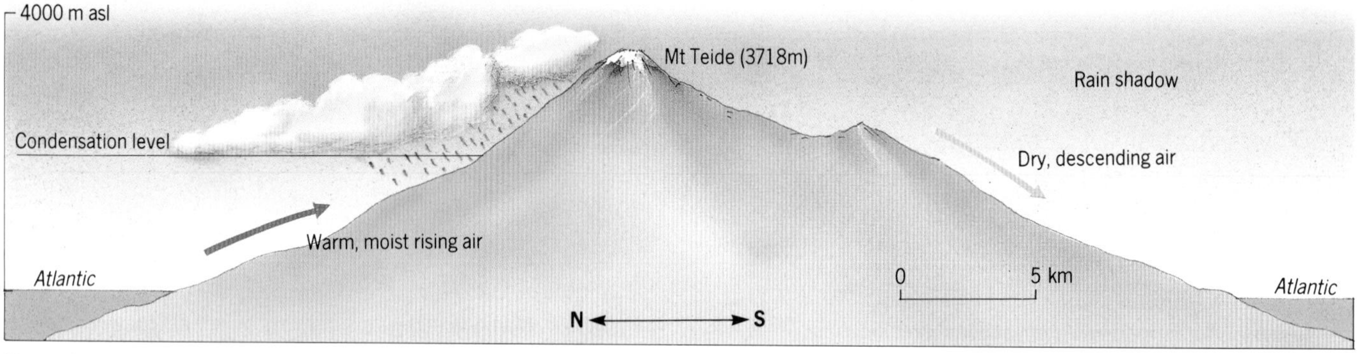

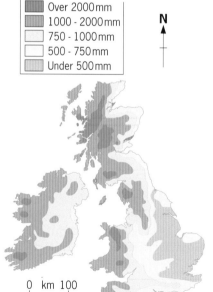

1500 - 4000m
200 - 1500m
0 - 200m

0 km 20

Punta del Hidalgo
Faro de Anaga
Bajamar
El Bailadero
Valle Guerra
Las Mercedes
Punta de Antequera
La Laguna
La Esperanza
San Andres
Santa Cruz de Tenerife
San Juan de la Rambla
Puerto de la Cruz
Garachico
Buenavista
Icod
Candelaria
Faro de Teno
Erjos
Arafo
Santiago del Teide
Guimar
Puerto de Guimar
Acantilado de los Gigantes
Araguayo
El Portillo
Chio
Pico Teide
Puerto de Santiago
Guia de Isora
Vilaflor
Loma de Arico
Adeje
Granadilla de Abona
N
Los Christianos
El Medano
Faro de la Rasca

SPAIN
20° 10°
CANARY ISLANDS
30°
Tenerife
AFRICA
0 km 500

4000 m asl

Mt Teide (3718m)

Rain shadow

Condensation level

Dry, descending air

Warm, moist rising air

Atlantic

0 5 km

N ←→ S

Atlantic

Figure 8.11 Orographic precipitation in Tenerife

Over 2000mm
1000 - 2000mm
750 - 1000mm
500 - 750mm
Under 500mm

N

0 km 100

Figure 8.13 British Isles: distribution of rainfall

evaporation and runoff; the rest infiltrates into the ground. Much of Tenerife's water supply is obtained from these groundwater reserves. However, there has been a steady increase in the demand for water. In fact, research shows that there has been a drop of about 150 m in the average level of the water table since the beginning of the twentieth century. This will obviously have consequences for human activities on the island.

Figure 8.12 Tenerife, south side: semi-arid landscape

4 Describe and explain the distribution of settlements on Tenerife (Fig. 8.9).

5 Tenerife is a popular European tourist destination area.

a You work for Tenerife's building authority and need to plan future hotels and tourist developments. Write a report to your manager explaining how the availability of water could influence the expansion of tourism on the island.

b In your report suggest what steps the authority could take to increase the amounts of water available.

6 Use an atlas to describe and account for the relationship between relief and precipitation as shown in Figure 8.13.

Cyclonic or frontal rainfall

Cyclonic or frontal rainfall is particularly important in the mid-latitudes, about 40–60° north and south. Low pressure systems, or depressions, moving from west to east, influence the climate of these zones. Where warm air is forced to rise over cold, cloud development and rainfall can occur (Figs 8.14–8.15). There is a fuller treatment of depressions and the weather they bring in Chapter 9.

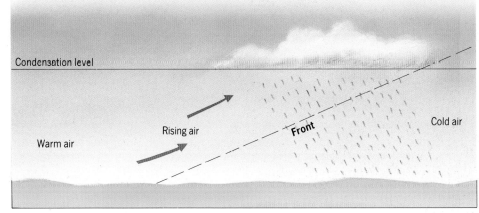

Figure 8.14 The formation of frontal precipitation

Storm damage report

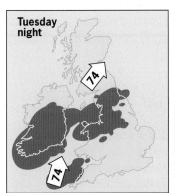

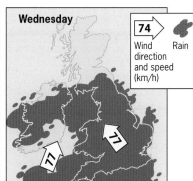

Figure 8.15 The results of heavy cyclonic rainfall (Source: The Guardian, 3 Dec. 1992)

7a Using Figure 8.15, suggest a variety of economic activites which may be affected by excessive cyclonic precipitation.

b What impact may this precipitation have on the economic activities?

Rainfall intensity

The duration and frequency of precipitation provide an important insight into the nature of storms in different parts of the world. This also has implications for the behaviour of rivers and the ecology of an area (Table 8.1, see section 4.4).

Table 8.1 Normal rainfall rates for the British Isles (below 300m asl) (Source: Boucher, 1974)

Continuous cyclonic rainfall		Shower/thunderstorm (30 minutes duration)	
0.3–1.0 mm per hour	Drizzle	0.3– 2.0 mm	Light shower
1.0–2.0 mm per hour	Fine rain	2.0– 4.0 mm	Moderate shower
2.0–3.0 mm per hour	Moderate rain	4.0–12.0 mm	Heavy shower
3.0–5.0 mm per hour	Heavy rain	12.0–25.0 mm	Heavy thunderstorm
Over 5.0 mm per hour	Very heavy rain	25.0–50.0 mm	Violent thunderstorm

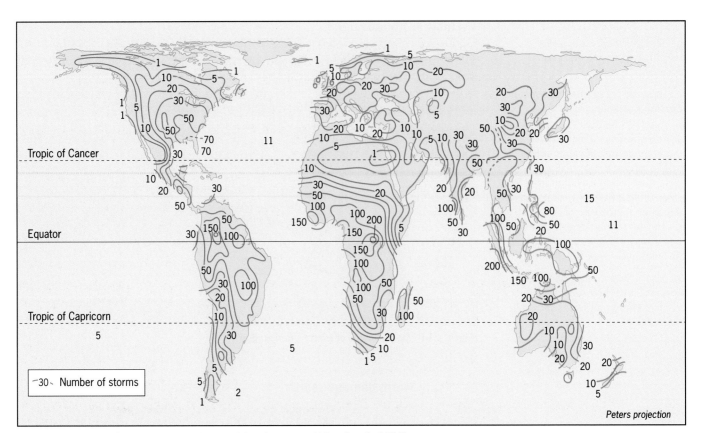

Figure 8.16 Mean annual global distribution of thunderstorms (*After:* Lamb, 1982)

?

8 Explain the global distribution of thunderstorms (Fig. 8.16).

The highest rainfall intensities in the world are recorded mostly in the tropics where temperatures and humidities are high. Intense **insolation** (see section 6.3) causes high rates of evaporation and strong convectional uplift. In addition, the convergence of the Trade winds near the equator aids this process. Although the reporting of tropical storms is not as reliable as in other parts of the world, we know that large parts of the Amazon and Zaire experience thunderstorms for over 100 days each year (Fig. 8.16). Orographic influences also encourage storm activity and further uplift. For example, storm frequencies of over 200 days per year have been recorded in western Java. The characteristics of rainfall in the tropics are examined more fully in Chapter 10.

Other forms of precipitation

There are a number of other precipitation forms which are important in their own context (Figs 8.17–8.18). These are described in Table 8.2.

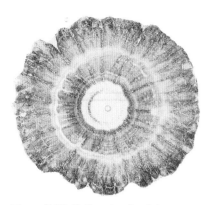

Figure 8.17 Hail: water droplets freeze to a cloud droplet, building layers until the stone becomes so heavy it falls to the ground. The largest hailstone recorded in the British Isles weighed 141 g and fell at Horsham, West Sussex in 1958. The world's largest recorded stone fell in Kansas, USA and weighed 758 g with a diameter of 10.9 cm.

More snow and very cold

The London Weather Centre said last night that more of the snow that blanketed eastern England yesterday was on its way, with temperatures today barely above freezing.

Police and motoring organisations urged commuters to allow more time for journeys and to reduce speed.

Up to six inches of snow fell, mostly over the eastern half of Britain from Scotland to the south. There were lesser falls in the Midlands and Wales and an inch in London.

Last night snowploughs fought to keep open the roads between England and Scotland. Blizzards swept over the Pennines, threatening to cut off moorland villages. Cars became stuck in five inches of snow on the A93 near Ballater, Grampian.

© Telegraph plc, London 1993

Figure 8.18 Snow: blizzards reported by the press (*Source: The Daily Telegraph*, 22 Nov. 1993)

Table 8.2 Types of precipitation

Precipitation	Cause	Example/occurence	Characteristics
Rain	• Convection: strong, upward-moving air • Orographic: rising warm, moist air. • Cyclonic: depressions rising over colder air.	• High temperatures • Upland areas • Mid latitudes	• Short-lived showers with thunderstorms • Steady rain/drizzle • Consistent rain
Drizzle	When turbulence produces low altitude stratus or stratocumulus clouds with little vertical extent (prohibiting development of larger droplets): moist air below is needed to prevent evaporation.	Upland areas e.g. western side of the British Isles	Tiny droplets less than 0.5 mm in size
Hail	Cumulonimbus clouds, approx 1000 x 14 000 m, produced by turbulence and convection. As a hailstone falls in a cloud it picks up small droplets that freeze to it producing a layer of clear ice (Fig. 8.17).	British Isles: hail storms often occur in the spring associated with westerly and northerly air streams. Mid-continental interiors: during afternoons after maximum heating and when unstable conditions prevail e.g. central and southern Asia, USA, central and southern Europe, South Africa and Australia.	Vary in appearance and size from 5 to 50 mm
Snow	Cool, calm conditions: stratus or altostratus clouds with temperatures well below freezing.	British Isles: mostly during winter when • a warm occluded front passes, • a depression passes just south of area of snowfall, • a showery northerly brings snow from the sea.	Supercooled water droplets freeze on tiny ice crystals
Sleet	Forms in warmer air than snow (3–4°C).		Sleet is a mixture of snow and rain, or partially melted snow: if rain or sleet passes through very cold air near the ground it forms a sheet of glazed frost (black ice).
Fog	Condensation of moisture in the air due to prolonged cooling from calm air, clear skies and long nights.		Water droplets suspended in air: freezing fog forms as supercooled water droplets exist near the ground below freezing point.
Fog drip	When low banks of coastal fog move inland and drift through forests: moisture condenses on leaves and drops to the ground.	USA: north Californian coast	50% of northern California's coastal precipitation is due to fog drip. Some vegetation would not exist without the extra moisture e.g. redwoods
Dew	Moisture formed as temperatures drop below condensation point at night.		Important source of moisture for plants and animals especially in arid environments. Freezing dew forms hoar frost with 'feathery' crystals.

9 If weather of the type described in Figure 8.18 was forecast, comment on how the following organisations would respond over the next few hours:
• police,
• motoring organisations e.g. AA and RAC,
• local authorities.

Fog

Fog is a mass of tiny water droplets suspended in the air. It follows that ideal conditions for fog are calm air, clear skies and long nights, when prolonged cooling will lead to condensation of moisture in the air. In the British Isles, most fogs occur in the autumn and mid-winter. There are two main types of fog. In inland areas, the most common type is **radiation fog**. This forms at night when the ground cools as heat is lost into the atmosphere due to radiation. Moisture in air near the ground will cool to the point where it condenses to form fog. Radiation fog is often patchy and tends to build up in valleys or depressions as colder air has a tendency to sink. Temperature inversions (see section 12.3) are associated with radiation fog.

Some coastal areas experience 'sea fog', which meteorologists call **advection fog**. This develops when warm, moist air passes over cooler sea, or a cold current, and reaches saturation point. Advection fog is common in spring and summer in the British Isles, and particularly affects the south-west and North Sea coasts. The Californian coast, near San Francisco, is also well known for its advection fogs (Fig. 8.19). Moisture in warm air passing eastwards over the Pacific condenses when it reaches the cold Californian Current. After dawn, fog is usually 'burnt off' as the sun's energy causes the water droplets to evaporate. The fog may, though, be more persistent in the winter.

The Meteorological Office defines fog as visibility less than 1000 m. This is appropriate for aviation but less applicable for other forms of transport and the general public (Table 8.3).

Figure 8.19 Advection fog: San Francisco

10 Attempt to classify precipitation forms into two categories: those generally associated with atmospheric stability and those associated with instability.

11 Use the national newspapers on a CD ROM and enter as key words different precipitation forms e.g. hail, snow and fog. Find recent examples of how human activities have been affected for each major form of precipitation.

Table 8.3 Difficulties for transport caused by various categories of poor visibility (*Source:* Met. Office)

Fog level	Effect on transport	Visibility
Dense fog	Severe disruption to most transport	Zero
Thick fog	Road, rail and aircraft on the ground delayed	50 m
Aviation fog	All aircraft landings affected to some extent	200 m
Mist or haze	Shipping and light aircraft affected	1000–5000 m

8.4 Precipitation and land use

Extremes of precipitation e.g. an excess or deficit of moisture, or a very marked seasonality, can have a noticeable effect on the landscape. In particular, they influence the distribution of vegetation. Precipitation is, in fact, one of the most significant elements of climate in determining patterns of agricultural land use. Such an influence of surplus water on land uses is typified by the **monsoon** in south-east Asia (see section 10.5).

Precipitation and land use in West Africa

Climatic background

Temperatures in West Africa are always high so the main factor limiting the periods of plant growth is the availability of water. The *inputs* of precipitation and *outputs* through evaporation vary considerably in the region both temporally and spatially (Figs 8.20–8.21).

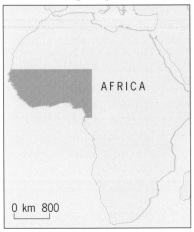

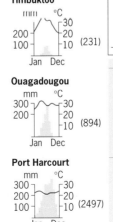

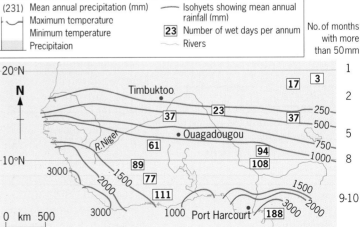

Figure 8.20 West Africa: spatial variations in precipitation

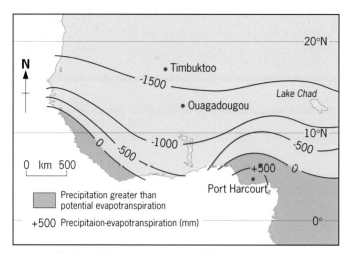

Figure 8.21 West Africa: mean annual difference between precipitation and potential evapotranspiration

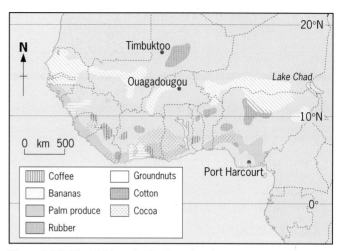

Figure 8.23 West Africa: distribution of major export crops

The persistently high temperatures in West Africa reflect the area's proximity to the equator and the sun's position almost vertically overhead at midday throughout the year. Precipitation is more complex to explain, though. The traditional view is that a 'belt' of rainfall moves towards the tropics in the summer months following the apparent movement of the overhead sun. This rainfall is associated with converging air currents and strong uplift at the Inter-Tropical Discontinuity (ITD) (see section 6.3). Although such frontal conditions are rare in the tropics (see section 10.2), the ITD separates the moist south-westerlies from the dry north-easterlies and thus brings rainfall to West Africa. Its movement is limited to north of the equator, moving out over the Atlantic during the winter months and to its northernmost position in August.

Effects on natural vegetation

The uneven distribution of rainfall and evaporation strongly influences the natural vegetation of the region

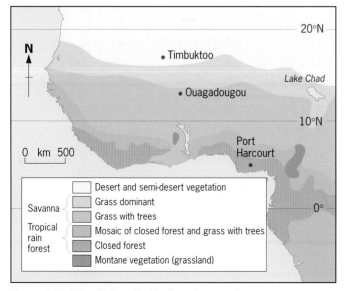

Figure 8.22 West Africa: distribution of vegetation types

(Fig. 8.22). Towards the edge of the Sahara there are many succulents and annual plant species, and all plants tend to have extensive root systems to obtain maximum moisture. Plants adapted to arid conditions are called *xerophytes*.

Effects on agriculture

Precipitation patterns also affect commercial and subsistence agriculture (Figs 8.23–8.25). The decline in rainfall totals and the length of the growing season particularly influences farmers' selection of arable or pastoral farming, although much traditional farming is mixed. The choice of planting date and of a suitable crop with an appropriate length of growing season is also critical (see section 6.4). However, rains are often late and unreliable, particularly inland, and crop failure is common. Further north towards the desert, crop production is only possible with the aid of irrigation. The warmth and long sunny days are ideal for crops, but high rates of evaporation lead to *salinisation* in the soil. This occurs when insolation encourages capillary movement of moisture upwards in the soil, bringing mineral salts to the surface.

12 The annual precipitation value for Ouagadougou is 894 mm. Use Figure 8.20 to estimate the total annual potential evapotranspiration.

13 In southern Nigeria where precipitation exceeds potential evapotranspiration (Fig. 8.21), what happens to the water that is not used in **evapotranspiration** or stored in the soil?

14 Use Figures 8.20 and 8.21 to describe the main changes in climate that occur along a transect from the coast near Port Harcourt to the north.

15 Describe and account for the pattern of natural vegetation in West Africa (Fig. 8.22) in relation to the climate (Figs. 8.20 and 8.21).

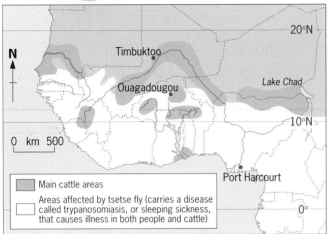

West Africa

Figure 8.24 West Africa: distribution of cattle

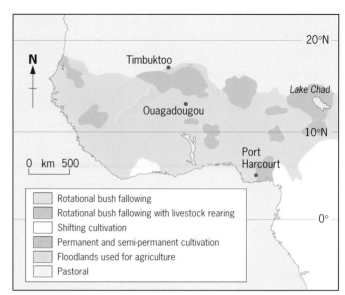

Figure 8.25 West Africa: dominant land use systems of peasant agriculture

16 State which two crops in Figure 8.23 seem to be most dependent on a high and well distributed rainfall.

17 Describe and suggest reasons for the distribution of cattle in West Africa (Fig. 8.24).

18 Suggest how the following factors may also have influenced the distribution of farming activities in West Africa:
a Communications and trade.
b The distribution of the tsetse fly.
c Soil quality and risk of soil erosion.

8.5 Acid rain

Robert Smith first drew attention to acid rain in Manchester in 1852, but his findings were largely ignored. Acid rain is an environmental problem that received considerable attention during the 1970s and 1980s when people became increasingly concerned about the effect acid rain was having on trees and life in lakes and the accelerated weathering to buildings and other structures.

The pH scale

Acidity or alkalinity is measured on the pH scale (Fig. 8.26). This measures the concentration of hydrogen ions in a substance. For example, if in pure water one part in one million (10^{-6}) was disassociated into hydrogen ions, then it would have a pH of 6. Note that the scale is logarithmic and therefore one step on the scale means a tenfold change in acidity or alkalinity.

19 Study Figure 8.26. Estimate how many times more acid pure rain and vinegar (similar in acidity to much acid rain) are in comparison with neutral or distilled water.

Figure 8.26 The pH scale

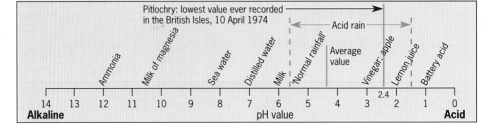

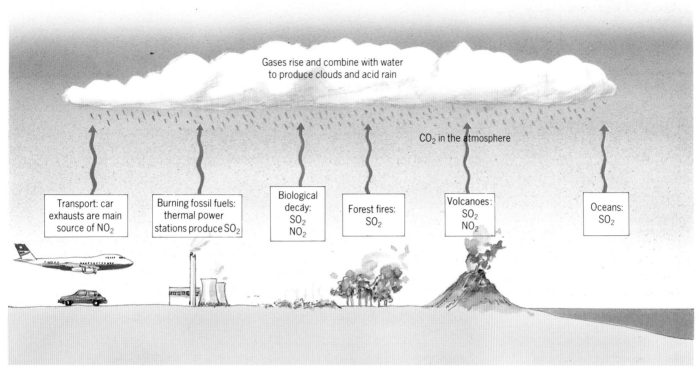

Figure 8.27 The formation and sources of acid rain

The formation of acid rain

We must not forget that rainwater is naturally acidic. It forms a weak solution of carbonic acid (H_2CO_3) resulting from carbon dioxide that dissolves naturally in water in the atmosphere. There are natural sources of sulphur dioxide (SO_2), like volcanoes and forest fires, that can also increase the acidity of precipitation (Fig. 8.27). The term acid rain generally refers to the acidity of precipitation resulting from pollution. This represents a disturbance to the system and includes *wet deposition* in forms like rain, snow, sleet and fog and the *dry deposition* of gases and particulate matter.

The consequences of acid rain

An international problem

It is largely the industrialised nations that are responsible for acid rain (Fig. 8.29). It is not, though, always the polluters that suffer from the problem. Increasingly, high chimneys eject pollutants into the upper **troposphere**. These can then travel many thousands of kilometres before falling to the surface. For example, the main destination area for pollution from central European countries is Scandanavia. Acid rain has even affected the tropical rainforests in South America. In cold climates, the pollution that falls in snow during the winter is released in the spring when the thaw occurs. This can lead to sudden changes in the acidity of lakes and rivers.

Vegetation and soil

In the 1980s much attention was given to damage caused by acid rain to trees. For example, many coniferous species were dying in European countries like Norway and Sweden. There is evidence, though, that this is still a problem (Fig. 8.28). The removal of natural bases from the rooting zone by acid rain also harms soil. As a result, the soils become more acid and levels of toxic substances increase. This not only affects natural vegetation but also agricultural productivity. Once acid rain enters the soil it affects soil chemistry by releasing excess quantities of substances like aluminium. These are harmful to trees as they affect the roots and prevent proper absorption of water. Trees are also stressed and become more susceptible to bacteria and disease.

Europe's trees deteriorating

European forests are continuing to deteriorate, mainly because of acid rain, the European Commission said yesterday. The Commission's annual forest damage survey showed that more than one in every five trees was damaged. Some 83 000 trees were tested. 'The report shows that there is an overall tendency towards a worsening of the forest condition in Europe.'

The Commission said coniferous forests were suffering the worst damage in Germany, Britain, Bulgaria, Czechoslovakia and Poland. Deciduous damage was worst among Danish beeches, Swedish birches and Portuguese oaks.

Figure 8.28 The effect of acid rain on trees (*Source: The Independent*, 23 Dec. 1993)

Figure 8.29 Acid rain: the global pattern

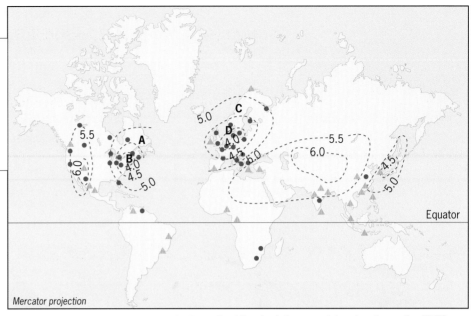

A Eastern Canada   C Southern Scandinavia
B North-east USA   D Northern Europe

~5.5~ Lines of equal pH rainfall (natural rain has pH of 5.6)

● Reported damage from acid rain (rain with pH of less than 4.5)

▲ Possible future damage from acid rain

Mercator projection

20a Suggest reasons for A, B, C and D being the main areas affected by acid rain in Figure 8.29.
b What do the areas labelled 'possible future damage from acid rain' have in common that might account for this?

21a Use the data in Table 8.4 to construct four pie diagrams showing the sources of acid rain for each of the Scandinavian countries. These should be drawn in proportion to the amount of acid rain received by each country in relation to the total for the receiving countries.
b Which four countries are the main sources of acid rain for Scandinavia?
c Comment on the extent to which your completed pie diagrams show the information more effectively than Table 8.4.

Table 8.4 Acid rain in Scandinavia: budget of oxidised sulphur, provisional estimate for 1989 (*Source:* DoE)

Source countries	Recipient countries Denmark	Finland	Norway	Sweden	Total
Belgium	1	1	2	3	7
CIS	0	30	5	12	47
Czechoslovakia	2	4	3	10	19
Denmark	13	2	3	10	28
Finland	0	42	1	6	49
France	2	2	3	4	11
Germany	14	16	18	45	93
Netherlands	1	1	1	2	5
Norway	0	1	8	3	12
Poland	3	11	6	22	42
Sweden	1	5	2	25	33
United Kingdom	9	7	30	22	68
Other known sources	4	7	13	16	40
Unattributable	6	40	45	48	139
Total countries	56	169	140	228	593

Amounts are measured in thousands of tonnes deposited per year

Lakes, water supply and buildings

High levels of lake acidity have led to the death of many species of fish in Scandinavian countries, e.g. salmon started to disappear from lakes in southern Norway shortly after World War 2. By the 1980s they had disappeared from about 2000 lakes in the country. In 1993 thousands of lakes in Sweden showed heavy fish losses while about 400 were virtually lifeless.

It is not the acid rain itself that kills fish but the aluminium washed into the lakes from the surrounding soil. Aluminium deposited in a mucus on the gills of fish inhibits the normal intake of salt and oxygen needed for survival. Fish gasp for breath, and as the salt content in their bodies drops, proteins are destroyed and the fish die. This changes the whole ecology of lakes. Crystal clear water becomes contaminated with algae, and a lot of organic debris collects around the edges that would normally decompose rapidly.

Acid rain also attacks buildings and, particularly through the process of *carbonation*, weathers limestone. Many of Europe's major cathedrals, including St Paul's in London, have been seriously damaged by acid rain (Fig. 8.30). Buildings like the Taj Mahal in India and the Parthenon in Athens have been affected.

Figure 8.30 Acid rain damage to Rheims Cathedral

Acid rain damage in the British Isles

In the 1960s and 1970s Britain earned the nickname 'the dirty man of Europe'. The Germans, Swedes and Norwegians later protested to Mrs Thatcher about Britain's huge acid rain emissions (still amongst the largest in Europe). However, during the 1980s media attention turned towards other concerns like ozone depletion and **global warming** and so acid rain became something of a 'Cinderella issue'. Now, in the 1990s, attention to acid rain is returning for several reasons. Firstly, there is the question of Britain's energy needs and reviews of the future of the coal industry. There has also been pressure from the UN Economic Commission for Europe Convention on Long-Range Transboundary Air Pollution and Britain's own pollution control initiatives. In addition, there is increasing evidence that acid rain is still causing serious environmental damage (Fig. 8.31).

Acid rain causes damage to British wildlife

THE headwaters of the River Severn are so acidified by airborne pollution from distant power stations, industry and vehicles that they are unable to nurture the shoals of minnows and sizeable brown trout that once were abundant here. The upper River Severn has no fish...

Hill streams and lakes across northern England have acidified, too...

A report by English Nature, in collaboration with the Countryside Council for Wales and Scottish Natural Heritage, shows more than 430 000 hectares of land protected as Sites of Special Scientific Interest (SSSI) – our most important wildlife locations – in England, Scotland and Wales are suffering acidification damage. This represents almost a quarter of Britain's SSSI area.

In the most severely affected region, north Wales, 57 per cent of the SSSI area – more than 48 000 hectares of land – is degraded or highly likely to have been degraded by a cocktail of sulphur dioxide and nitrogen oxides, which can, on occasion, transform pure rainwater into a brew more akin to vinegar.

As a result, many of the plants and animals on the rugged moors and bogs are in decline or have already disappeared. Conservationists are concerned for rare fish species such as powan and arctic char; for lake and stream populations of trout, salmon and minnows; for frogs; elusive otters; birds such as dippers, whose survival depends on healthy populations of water-living insects; and for the fragile crust of lichens and mosses that clothe the peat bogs over vast tracts of mountain and moor. It is a depressing list.

The largest amount of pollutant that will not cause chemical effects in the soil leading to long-term ecological change is known as the 'critical load'. Maps of soil critical loads produced by the Department of the Environment's Critical Loads Advisory Group show that more than 108 000 km^2, chiefly in Wales, Scotland, Cumbria, and northern Pennines, Dartmoor, Bodmin Moor and the New Forest, exceed their critical load. Their capacity to neutralise any more acid rain falling on them is already exhausted, so ecological change is unavoidable.

Figure 8.31 Research reveals the extent of acid rain damage to wildlife in Britain (*Source: The Independent*, 8 June 1992)

Solutions to the problem

We can tackle the problems created by acid rain by reducing the causes (inputs), or by controlling the effects of the rain in the environment (outputs). There are basically two approaches to each of these.

Technological solutions

People have tried adding lime to lakes to reduce acidity. However, this only tackles the symptoms (outputs) of acidity and not the cause (inputs). The technology to control the cause of acid rain is widely available but it is expensive. For example, sulphur can be removed from coal and oil before it is burnt. Alternatively, power stations and industry could use coal with a low sulphur content. They can also be fitted with new burners that inject limestone to remove the remaining sulphur dioxide and/or be fitted with a filter attached to the chimney.

Collectively, these measures could dramatically reduce emissions (Fig. 8.32). Ultimately, developing clean alternative energy sources like wind and wave power will also help solve the problem.

Political solutions

Acid rain is an international problem that therefore requires international co-operation and initiatives to reduce emissions. The first major step came in 1979 when the United Nations set up the Convention on Long Range Transboundary Air Pollutants. The UN modified this in 1983 and fifteen European countries plus Canada signed and agreed to cut their sulphur emissions by 30 per cent by 1993 from 1980 levels. Some nations are now seeking an 80 per cent reduction following a further UN initiative started in Geneva in 1993.

In June 1988 the UK agreed the Large Combustion Plants Directive with the other EC countries. This Directive sets out emission limits for SO_2, NO_2 and dust from new combustion plants. It also committed the UK to reducing emissions of SO_2 by 60 per cent and NO_2 by 30

British Isles

per cent from the 1980 level by 2003 from exisiting large combustion plants.

Some countries have cut their SO_2 emissions since 1980, e.g. Belgium (50 per cent), the Netherlands (50 per cent) and France (60 per cent), but the UK has only achieved about a 25 per cent reduction. Although the problem of acid rain is being tackled in areas like Western Europe, the problem is now expanding to the newly industrialising countries where emissions could rise by several hundred per cent. It remains to be seen whether they will have the political will to control pollution levels.

BRITISH negotiators became resigned to asking the Cabinet to reopen the Coal Review after Britain's offer of 70 per cent reductions in sulphur pollution by 2005 were rejected by Norway, Sweden, the Netherlands and Germany.

Yesterday, when European countries announced their targets for the first time in the two-year negotiations, Britain was found to be offering a lower reduction than Slovenia, part of the former Yugoslavia, and the same as Poland, one of the most polluted countries in the former communist bloc. Computer models show that Britain, said to export more acid rain than any country other than Russia, would have to cut its sulphur emissions by 76 per cent to reach the standards set in the UN negotiations.

Independent estimates show that Britain would have to fit flue gas desulphurisation equipment costing 400 million pounds or burn low sulphur coal to meet the target.

© Telegraph plc, London, 1993

Figure 8.32
£400 million pollution controls demanded
(*Source: The Daily Telegraph*, 1 Sept. 1993)

22 Study Figures 8.28–8.31. To what extent is acid rain still a problem in the 1990s?

23 Study Figure 8.32 and identify the main obstacles to controlling acid rain emissions in Britain.

24a What different values are reflected in Figure 8.33?
b Give your own suggestions to solve these opposing positions.

25 Suggest what steps could be taken at various levels to reduce the problem of acid rain. Use the following headings.

a Government, national and local
b Industry, manufacturing and building
c Individuals

26 Why might newly industrialising countries like Brazil be less willing and able to control the problem of acid rain?

27 Essay: Choose an area of continental size, such as North America or Europe, and account for variations in the incidence and spatial distribution of acid rain. Evaluate the steps that can be taken to reduce the effects of acid rain.

PRESENT traffic growth cannot be sustained if public health and the countryside are to be protected and pollution of the atmosphere prevented over the next 20 years, Mr Gummer, Environment Secretary, said yesterday. Launching a consultation paper on the kind of Britain people would like in 2012, he called for 'a major national debate' on damaging trends such as increasing use of the car, said by his paper to cause 'perhaps the greatest impact on the environment'.

The timing of his green paper was seen as a challenge to the Transport Secretary, Mr MacGregor, who is expected today to announce a public inquiry into the widening of the Staines section of the M25 to a 14-lane super-highway, the largest outside North America.

At the weekend it was disclosed that other motorway bridges are being widened so that motorways elsewhere can be increased to twelve lanes.

© Telegraph plc, London, 1993

Figure 8.33 Cars put us on the road to ruin (*Source: The Daily Telegraph*, 22 July 1993)

Summary

- Condensation does not necessarily lead to precipitation. Rain droplets form around minute particles called condensation nuclei.
- There are two major theories of precipitation formation, the Bergeron–Findeison process and the coalescence theory. These have been expanded as a result of recent research and the availability of new data.
- The artificial process of cloud seeding is based on the Bergeron–Findeison theory of rain formation.
- Clouds and precipitation develop when a mechanism such as relief, convection or a front encourages air to rise to the condensation level.
- Precipitation forms can be classified into two groups. Major forms account for large amounts of moisture reaching the ground while minor forms account for less moisture and are not so widespread.
- Precipitation extremes can have a pronounced effect on the landscape, particularly influencing vegetation distribution. This is also one of the most significant elements of climate in determining patterns of agricultural land use.
- Rainwater is naturally acidic. It forms a weak solution of carbonic acid resulting from carbon dioxide that dissolves naturally in water in the atmosphere.
- The term acid rain refers to the acidity of precipitation that results from pollution.
- Acid rain is an environmental problem that received considerable attention during the 1970s and 1980s. Many recent efforts have reduced the problem in economically developed countries.
- Acid rain remains a problem in both economically developed and economically developing areas and is growing in the newly industrialising countries.

9 Mid-latitude systems

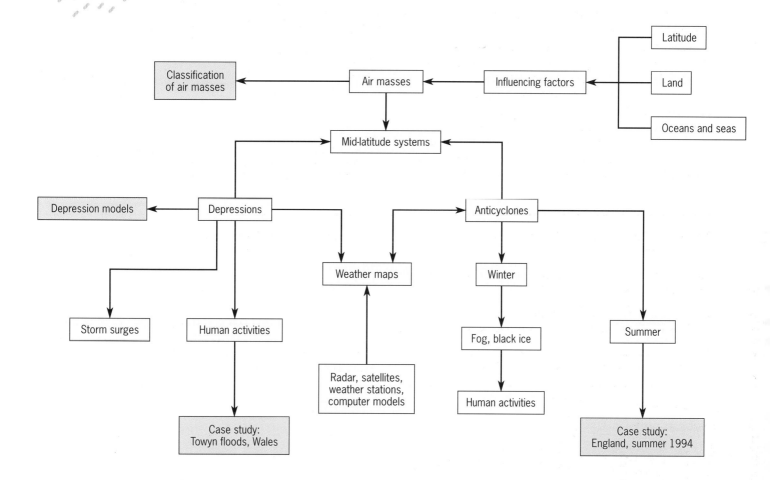

9.1 Introduction

The greatest variability in the atmosphere occurs in the middle latitudes (Fig. 9.1). The atmosphere in these latitudes is characterised by strong vertical changes in temperature (thermal gradients). The westerly air currents are the single most important feature of the mid-latitude circulation. In the upper atmosphere they are at their most powerful in the polar front **jet stream** at the base of the **stratosphere** (see section 6.5). They are largely responsible for the continuous sequence of low **pressure** systems (**cyclones** or **depressions**) and high pressure systems (**anticyclones**) that fundamentally affect the **weather** and **climate** at these latitudes.

We will examine the origin of these low and high pressure systems and how they affect our weather and climate. We will assess the extent to which depressions can exert a powerful influence on human activities, even leading to hazardous conditions (Figs 9.2 and 9.3). We have seen already (see section 6.5) that changes in the position of the westerlies and particularly the jet stream can lead to intense depressions with powerful winds and heavy rainfall. Contrasting anticyclonic conditions can also affect people in various ways. In the summer, periods of prolonged dry weather may seem very pleasant but again they can create difficulties. In addition, anticyclonic conditions in the winter result in very cold, dry conditions that can lead to hazardous conditions for drivers and often the loss of life.

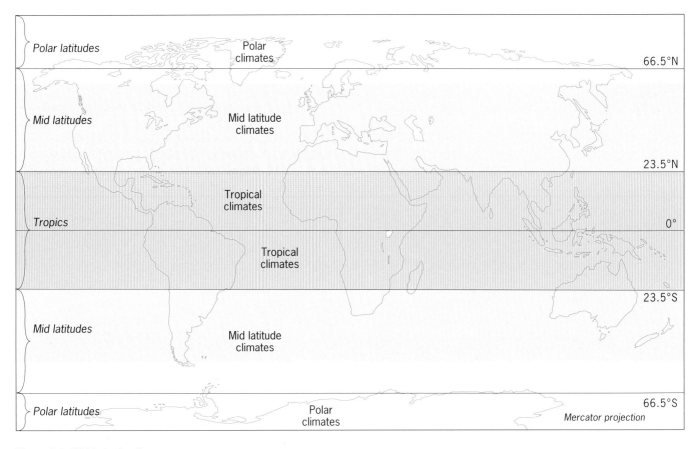

Figure 9.1 Mid-latitude climate zones
(and other major latitudinal zones)

Figure 9.2 The effects
of extreme weather
associated with
depressions (*Source:
The Independent*, 7 Nov.
1994)

32 die in storms

At least 32 people were killed yesterday when the worst rainstorms to hit north-west Italy in 80 years turned the area into a disaster zone. At least 12 other people were missing in the Piedmont region bordering France where more than 60 hours of torrential rain cut off towns and villages, shut main roads and railways and left hundreds of people without shelter.

At least 21 of the known dead were drowned, trapped in their cars in the Cuneo area of Piedmont, south of Turin. Other victims were buried by landslides or crushed to death when the force of water flattened their homes.

Five people were also missing in southern France and Corsica, swept away in their cars in floods that turned small rivers into torrents, French police said.

?

1 With reference to Figure 9.1 describe which parts of the world are referred to as the mid-latitudes.

2 Study Figures 9.2 and 9.3 and state the ways in which weather linked to depressions affected people in northern Italy.

Figure 9.3 Storm damage near Crescentino in the Piedmont region, Italy, 8 November 1984

9.2 Air masses

The air mostly takes its characteristics of temperature, momentum and **humidity** from the earth's surface over which it passes. It follows that if air remains more or less stationary for a period of time, it will take on the characteristics of the land underneath. Such an area of air that is largely uniform (with the same properties) is called an **air mass**. It follows that suitable locations for air masses to develop need to have a fairly uniform geographical character. Examples include the north Atlantic, the polar ice caps and large desert areas like the Sahara. Anticyclonic or stable atmospheric conditions will also encourage the development of air masses.

An air mass can extend over thousands of square kilometres, although the term can also refer to a more localised situation. Where two air masses with different properties meet, a sharp gradient in terms of the physical properties occurs called a **front** (see section 9.3). These appear on weather maps and often on weather forecasts on television and in newspapers (see Fig. 9.13).

Although the concept of air masses has been extensively used by geographers, and in the past by meteorologists, it does have a number of difficulties:

1 In reality there are no areas that are completely uniform geographically. Consequently there are variations of temperature and humidity within an air mass.
2 As air masses move from one area to another their properties change, so in practice gradients of the physical properties occur.
3 The idea of an air mass focuses attention on surface movements and properties of air. In reality, surface movements and pressure systems are connected with upper atmosphere processes (see section 6.5).
4 The idea of air masses is essentially descriptive. It is therefore not particularly useful in developing explanatory models of atmospheric processes.
5 Important processes such as **precipitation** are linked more to the dynamics of the atmosphere, like **instability** and convergence, rather than the characteristics of an air mass.

Despite these limitations, the air mass concept still provides us with a useful descriptive tool. Meteorologists still analyse air mass frequency and periodicity, while reference to air masses in weather forecasts gives us some indication of expected weather.

The classification of air masses

Air mass type (Lamb)	Air mass type	Season	Typical weather conditions (Lamb)
Northerly	mA	Winter	Cold, snow showers over north Britain and east coast. Blizzards with polar lows
		Summer	Cool, showers along the east coast
North-westerly	mA or mP	Winter	Cool, changeable, showery weather
		Summer	Cool, showers confined to western areas
Westerly	mP(warm) mT	Winter	Unsettled weather, mild spells, likely to be storms
		Summer	Fairly cool, blustery weather. Cloud at times
Southerly	cT(summer) cP mT	Winter	Mild and wet in the west. Usually drier and less mild in the east
		Summer	Warm with thundery tendency
Easterly	cP cA	Winter	Cold or very cold. Snow showers in the east
		Summer	Warm and dry
Cyclonic	mP or mP(warm) mT	Winter	Wet and windy
		Summer	Rather cool, cloudy sometimes thundery
Anticyclonic	mP(warm)	Winter	Cold, frosty. Fog in November/December
		Summer	Warm and dry

Table 9.1 Lamb's classification of air masses for the British Isles

Classification of air masses

One of the classic groupings of air mass types for the British Isles was developed by H H Lamb. Table 9.1 shows that this is based on a synoptic approach (see section 9.3) and emphasises different types of weather. Today, air masses are generally classified into *Polar* or *Tropical* and then sub-divided, depending on whether they have a *continental* or *maritime* source (Table 9.2). The corresponding abbreviations relate to the type of air mass, so that *mP* refers to *maritime polar* air and *cT* to *tropical continental* air. Figure 9.4 shows the areas of origin for air masses that affect the weather of the British Isles.

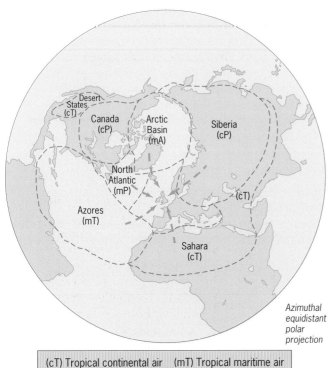

(cT) Tropical continental air   (mT) Tropical maritime air
(cP) Polar continental air   (mP) Polar maritime air
(cA) Arctic continental air   (mA) Arctic maritime air

Azimuthal equidistant polar projection

Figure 9.4 Air mass sources of the northern hemisphere and the British Isles

---

?

---

3 Give reasons why the Sahara and the polar areas are suitable for the development of air masses.

4 Explain why:
a Easterly (cP) air is associated with warm and dry weather in the summer.
b Southerly (mT) air is associated with mild, wet weather in the south-west.

5 Essay: Define the term air mass and explain the characteristics of the different air masses that affect the weather of the British Isles. Use Figure 9.4 and Tables 9.1 and 9.2 to help you.

Table 9.2 Major air masses and associated weather

Air mass	Characteristics and associated weather
Polar maritime (mP)	**Source:** North Atlantic near south Greenland.
	Weather: mP occurs in British Isles for about 40% of the year and the type of weather depends on the time of year. Although the air is originally cold, it warms while crossing the North Atlantic, becoming very unstable on reaching the British Isles. It produces cumulus clouds and often short but heavy showers. When approaching from the south-west, mP is drawn in by depressions curving over the Atlantic. This mPw (Polar maritime, warm) air cools as it approaches and becomes more stable, resulting in sea fog in summer in coastal areas with light cloud cover. The high moisture content may combine with strong heating to produce cumulus cloud and showers. In winter, low stratus cloud and dry conditions develop. Uplift over upland areas may encourage instability and rainfall, contributing to frontal rainfall.
Tropical maritime (mT)	**Source:** Azores high pressure system in the tropical North Atlantic (see Fig. 6.13).
	Weather: mT influences weather in British Isles for about 10% of the year, although it can occur in all seasons. Its tropical source causes the air and resulting weather to be warm and moist. The air tends to cool as it approaches the British Isles and therefore becomes more stable. This leads to sea fog and clouds with little vertical extent e.g. stratus. In summer, strong heating from land surfaces encourages warm moist air to rise by convection, leading to thunderstorms.
Continental air: Polar (cP) and Tropical (cT)	**Source:** cP originates from anticyclonic air over Scandinavia.
	Weather: cP is uncommon in the British Isles, occurring in winter and early spring. Dry, cold and stable air results in cold, dry weather and clear skies. Some moisture and warmth may be picked up as the air crosses the North Sea, bringing light snow showers to the east coast.
	Source: cT originates from north Africa.
	Weather: cT is the most infrequent in the British Isles – being warm and dry. Conditions are very stable with clear skies and heat waves in summer.
Arctic air	**Source:** Originates over frozen Arctic seas.
	Weather: Occasionally affects the British Isles in winter. As the air crosses the Atlantic, it picks up moisture and warmth (becoming mP air as it is not as cold as cP air). The air becomes unstable, bringing heavy snow and blizzards particularly to Scotland and the north of England.

9.3 Depressions

We have previously seen (see sections 6.3 and 6.4) that low pressure can occur at a variety of scales. At the mesoscale, low pressure systems can cover thousands of square kilometres, including the largely permanent systems that form part of the global pressure patterns. **Hurricanes** are also at the upper end of the scale, while at the other end intense low pressure phenomena like tornadoes occur.

Depressions are low pressure systems (lows) that occur in the middle latitudes and are associated with the meeting of cold and warm air masses. They were formerly called cyclones but this term should be avoided and reserved for the tropical systems that we refer to as hurricanes (see Chapter 11). Nevertheless meteorologists still refer to cyclonic rainfall in association with depressions (see section 8.3).

Depression models

The traditional model of a depression

We have previously seen that depressions are surface pressure systems linked to upper air movements that are part of the global atmospheric circulation and transfers of energy (see section 6.5). Figure 9.5 shows a typical depression that lies ahead of a **trough** in the upper westerlies. Depressions form at a semi-permanent boundary between cold polar air and warmer air from the tropics. This boundary is known as the polar front. A front is the boundary between two air masses and reflects the nature of the air *behind* rather than *forward* of the front.

Much of our early understanding of depressions came from the work of a group of Norwegian scientists working between 1918 and 1945. The Bjerknes model was widely accepted for many decades. Although new models have now largely replaced these ideas, they remain a useful and accurate approach to looking at the behaviour of a depression and the related weather as it passes over an area.

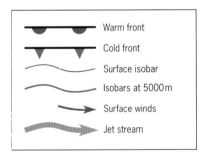

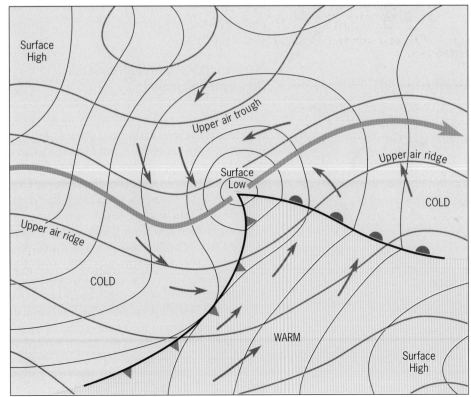

Figure 9.5 Upper and lower atmosphere features of a typical mid-latitude depression

Depression models

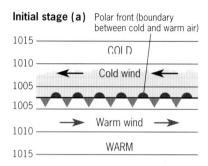

Initial stage (a) Polar front (boundary between cold and warm air)

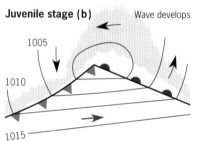

Juvenile stage (b) Wave develops

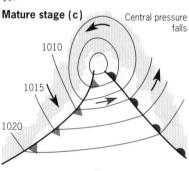

Mature stage (c) Central pressure falls

Occluded stage (d) Cold front has caught up with warm front

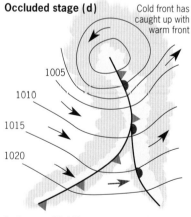

Each stage: 24–36 hours apart, travelling approximately 800 km

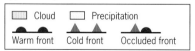

Cloud Precipitation
Warm front Cold front Occluded front

Figure 9.6 Stages in the development of a mid-latitude depression

Life history of depressions

Depressions grow from an early stage to maturity to eventually fill and die out (Fig. 9.6). From the preconditions in stage (a), an instability begins on the polar front in stage (b) as cold air pushes warm air north in the northern hemisphere. Because the warm air is less dense than cold air, it rises over the cold air. The boundary between these two air masses forms the warm front. In contrast, the colder advancing air to the west is more dense and so undercuts the warmer air. The boundary separating these two air masses forms the cold front.

As the depression develops in stage (c) the cold front tends to move faster than the warm front. This is because it is easier for cold air to force warm air upwards than for warm air to encourage cold air to sink. Eventually therefore, the cold front catches the warm front and the sector of warm air is lifted away from the surface. This is the **occluded** stage (d) and is marked by an occluded front on weather maps. By the time occlusion occurs, the warm air has moved north and is trapped on the poleward side of the polar front. This represents an important transfer of energy from low to high latitudes.

Characteristics of depressions

A marked comma-shaped cloud pattern that is typical of a mature depression is often distinct on satellite images (see Fig. 9.14). A mature depression can be up to 1700 km in diameter and can last for between four and seven days. Figure 9.7 shows the cloud types associated with the different parts of a mature depression and the precipitation related to each front.

Precipitation occurs in three parts of a depression: at the warm front where warm air rises over cold air; at the cold front where cold air undercuts the warm air; and in the centre of the depression where air rises to compensate for inward-moving air at ground level. In each case, cloud develops at the **condensation level**. Rainfall will then result if there is sufficient **condensation** and **hygroscopic particles.** The release of **latent heat** associated with condensation and cloud development also helps to drive the depression by encouraging uplift of air and intensifying lower pressure near the ground. Our understanding of depressions has grown rapidly since World War 2, largely due to research into the relationships between frontal processes, depressions and rainfall. Technological developments have made much of this possible.

Modern models of depressions
Conveyor belt model

Figure 9.8 shows the most widely accepted model of a depression often called the conveyor belt model based largely on work by Browning et al. (1973). At the 500 millibars (mb) level is the cold dry air of the upper **troposphere** which divides to pass above the depression and form the mass of cold air behind the cold front. A flow of cold air approaches the centre of the depression from the east or south-east originating from an area of high pressure. This initially passes parallel to the warm front and, as it leaves, rises in a clockwise direction out of the system at a higher altitude. Near the surface in the west is another conveyor belt several hundred kilometres wide and several kilometres deep. This is a belt of warm air originating from an area of high pressure to the south. As it approaches the cold air at the warm front, it is forced to rise in a clockwise direction out of the system. This warm air then flows away from the depression in the upper troposphere. This is an important mechanism by which energy is transferred from lower to higher altitudes and latitudes. In fact, depressions may be responsible for as much as 30 per cent of the poleward transfers of heat (see section 2.5). In a mature depression the uplifted air is carried away faster than the rate at which air arrives at the centre of the depression near the surface. This encourages the pressure at the centre to fall. The upper fast-moving air currents steer the depression and therefore influence its rate of movement and direction.

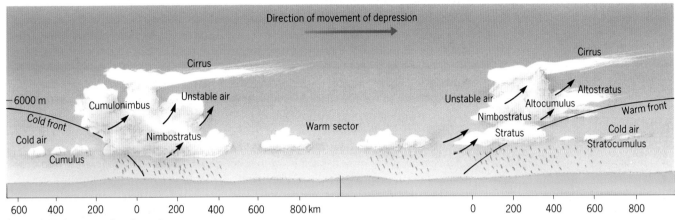

Figure 9.7 Cross section through a mature depression

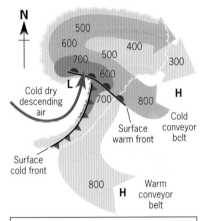

Figure 9.8 Conveyor belt model of a depression

6 Look ahead to Figure 9.14.
a Describe the position of the centre of the depression and the fronts. Explain your answer.
b Name the air masses most likely to be affecting the following areas:
- the south-west of the British Isles,
- north-west Scotland.

7 Describe in detail the sequence of weather that would occur as a depression passed over a point on the ground. Refer to:
a atmospheric pressure
b cloud cover
c wind speed and direction
d precipitation.

Ana and kata fronts

The fronts that form an important part of a depression are not lines but rather zones. Research in the 1960s showed that a front can be up to 500 m wide, 1 km above the ground. They are far more complex than originally thought and composed of cells of rising warm air and areas of very slowly descending cold air. This accounts for the patchiness and bands of rainfall that have been revealed by detailed research.

In the 1950s two types of front were identified. These have different types of air movement which in turn influence the nature of the frontal weather.

Ana fronts occur where warm air is generally rising in a flow that is not quite parallel to the cold front. This encourages instability and heavy rainfall (Fig. 9.9). Where the air flows away from the cold and warm fronts and is not forced to rise, a **kata front** develops. The broadly descending air leads to more stable conditions and less intense rainfall (Fig. 9.10).

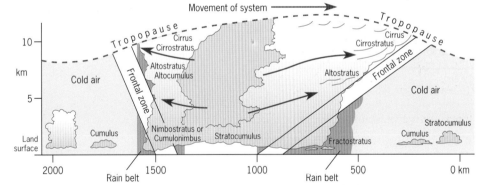

Figure 9.9 Ana front

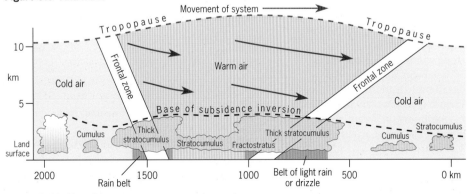

Figure 9.10 Kata front

8 Use the following observations to complete a station plot like the one in Figure 9.12.
Wind direction: north westerly
Wind speed: 8 to 12 knots
Cloud coverage: 6 oktas
Cloud type: stratocumulus
Precipitation: intermittent heavy rain
Temperature: 17°C
Pressure: 1014 mb

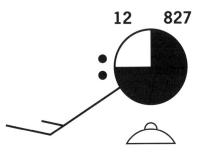

Figure 9.12 A typical station plot for a synoptic weather map

Pressure readings in millibars are nearly always in the upper 900s or lower 1000s. Some meteorological maps omit the 9 or 10 and express the rest of the figure in tenths of a millibar. Thus, 982.7 mb appears as 827; 1017.2 mb appears as 172. Depending on the size of the figure shown on the map, it should be obvious whether a 9 or a 10 has been omitted.

?

9 Study Figure 9.6 and the weather map in Figure 9.13. What stage has this depression reached?

10 Study Figures 9.13, 9.14 and 9.15.
a How does the weather map reflect the dense cloud cover shown on the satellite image?
b State two ways in which you could conclude from the weather map (Fig. 9.13) that there were strong winds in the British Isles.
c Use Figures 9.13 and 9.14 to describe and explain the weather in Norfolk as shown on Figure 9.15.
d Assuming the weather pattern moves eastwards at an average speed of 40 km/h, provide a weather forecast for Norfolk for the period from 0600 hours to 1400 hours in terms of precipitation, temperature, wind speed and direction.

Depressions and weather maps

Before we can look at how depressions are portrayed on weather maps, it is necessary to examine how the data are displayed. A **synoptic weather chart** or map shows the weather over an area for a particular time. As with an Ordnance Survey map, we use various symbols to display a wide range of information (Fig. 9.11). We need these symbols to help us interpret weather maps efficiently and Figure 9.12 shows how information would be displayed for a typical weather station. In this example there was a south-westerly wind of 13–17 knots. The cloud type was large cumulus and coverage six eighths, with intermittent moderate rain occurring. The figures indicate a temperature of 12°C and atmospheric pressure of 982.7 mb.

Once we have completed a station plot, we can then draw in the **isobars** and fronts. We can identify the position of fronts by looking for zones of sudden temperature (or **dew point** temperature) change or marked bends in the direction of isobars indicating a change in wind direction.

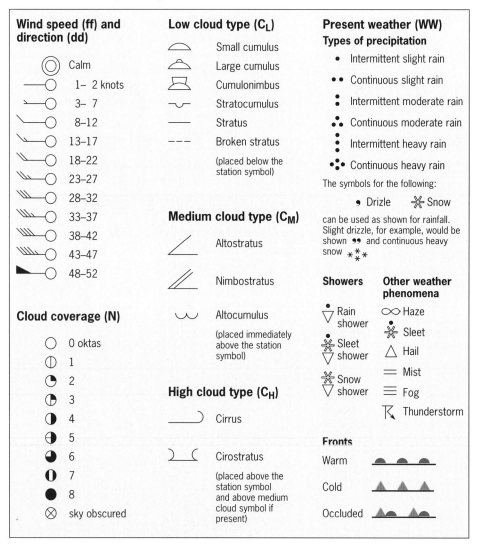

Figure 9.11 Symbols used on weather maps

We often associate depressions with strong winds and heavy rainfall. Figures 9.13–9.15 all show the same depression in different ways. The satellite image shows dense cloud cover over the British Isles. The comma shape is very clear on the satellite image, and the band of cloud crossing the north-west corners of France and Spain and the British Isles can be matched to the front on the weather map. Figure 9.16 is a brief summary of the weather on the weather map for a particular day.

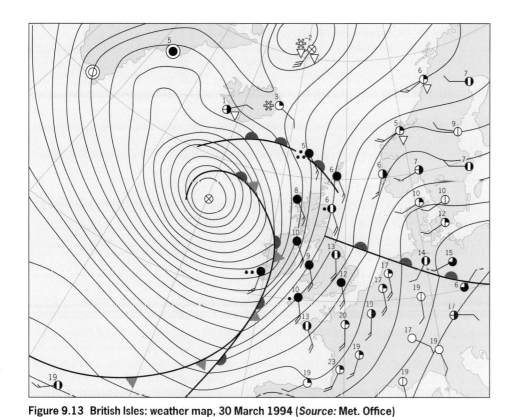

Figure 9.13 British Isles: weather map, 30 March 1994 (*Source:* Met. Office)

Many areas were windy, particularly in the west where there were gale and severe gale force winds. All areas had rain although this was heaviest in Scotland, northern Ireland and Wales. East Anglia and the South-East had a little rain and drizzle later in the afternoon and early evening. Much of England had a cloudy day. Most places had temperatures above average and parts of the South-East were warm.

Figure 9.16 Summary of a weather forecast for 30 March 1994

Figure 9.14 British Isles: satellite image for the same depression, 30 March 1994 (*Source:* Met. Office)

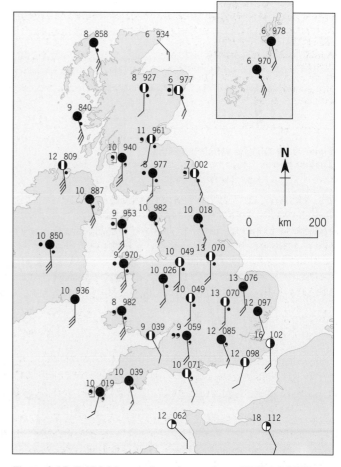

Figure 9.15 British Isles: surface observations, 30 March 1994 (*Source:* Met. Office)

9.4 Depressions and human activities

The powerful winds, heavy rainfall and low pressure that often occur with depressions frequently create difficult, and even hazardous conditions for people (see Figs 9.2–9.3).

Storm surges

Storm surges develop when low pressure causes the surface of the sea to rise slightly. This is known as the *inverse barometer effect*. If this combines with a high tide and strong onshore winds, water can pile up against the coast well above normal levels. Low-lying coastal areas are consequently flooded, causing widespread damage (see Fig. 11.8). This is not only a hazard in the mid-latitudes but also in the tropics when hurricanes occur (see section 11.4).

The east coast of England has experienced surges in 1825, 1894, 1906, 1916, 1953 and a minor surge in 1977. The 1953 surge is one of the most famous in recent history with widespread effects in England and the Netherlands. It resulted from a depression moving to the south-east which led to powerful north-westerly winds that built up to gale force. On 31 January, this depression combined with spring tides to cause a surge of sea water in the Straits of Dover and along the east coast (Fig. 9.17). The surge produced a tide level 2m above the normal spring tide level.

Fearing the worst effects of further storm surges on London, the river authorities built a tidal barrage across the Thames at Woolwich (Fig. 9.18). In fact, the threat of storm surges is likely to increase in the south of England for two reasons. Firstly, this part of the country is slowly sinking (isostatic adjustment) at a rate of about 1 mm a year as a response to the loss of weight when the Pleistocene ice sheets melted. Secondly, scientists believe that **global warming** could result in warmer seas that would intensify storms and encourage more powerful and frequent storm surges (see section 4.4).

Figure 9.18 Oblique aerial view of the Thames Barrier

Figure 9.17 Residents are rescued from floods on Canvey Island, Essex, 2 February 1953. Damage caused by the 1953 surge was dramatic: over 64 000 hectares of land were flooded in England, 141 000 hectares in the Netherlands; the physical landscape changed as sand dune complexes in England and the Netherlands were breached; cliffs at Herne Bay in Kent receded by 3 m.

Cyclonic conditions and the Towyn floods, Wales

In February 1990, exceptionally powerful winds and gales swept over much of England, leading to widespread flooding and damage. This affected the Thames Valley, the Severn Valley in Worcestershire and Gloucestershire and many parts of Wales. Estimates for insurance claims from storms by the end of February 1990 were put at £2.5 billion – about twice the normal figure for any given year.

By far the worst affected area was Towyn on the north coast of Wales. Waves in the Irish Sea were driven onshore by the powerful winds, resulting in a storm surge. The old Victorian sea walls along the coast were breached and the low-lying coastal plain was covered with sea water up to 4 km inland from the coast. The entire village of Towyn was inundated and thousands of people had to leave their homes (Fig. 9.19). Apart from the disruption to people's lives, there was extensive damage to property and disruption to communications (Fig. 9.20). The railway line along the north coast was out of action for several days and many main roads were impassable in the area. About 100 people had to be taken to hospital for treatment and to recover from shock. Many months after the floods, agricultural land and gardens were still recovering from the salt water. It was estimated that about £32 million would have to be spent on new sea defences to prevent floods of this magnitude occurring in the future.

Figure 9.20 Flood damage at Towyn, North Wales, February 1990

The windy weather and floods of early 1990 were exceptional. This led many people to believe that global warming was to blame. However, according to the Meteorological Office, a few severe storms during a winter, even within one month, are unusual but not unprecedented.

All hands to the Towyn pump

Police, firemen, ambulance crews and lifeboat men, using inflatable inshore rescue craft, last night helped more than 2000 people living along 40 km of the North Wales coast in the Towyn and Kimmel Bay areas to flee as the midnight high tide approached.

Helicopters, which had airlifted 37 elderly and infirm people to safety as flood waters rose to 1.5 m in Towyn during the afternoon, later hovered over the town as the tide turned at around 12.15 am.

Earlier, fire appliances had been used to ferry people suffering from shock and hypothermia through waist-deep water out of their houses. Electricity was cut off and telephone lines were down.

One woman was rescued from her home with her two day old premature baby, and 200 children had to leave a primary school when water poured in.

Figure 9.19 Impact of the Towyn floods, February 1990 (*Source: The Guardian*, 27 February, 1990)

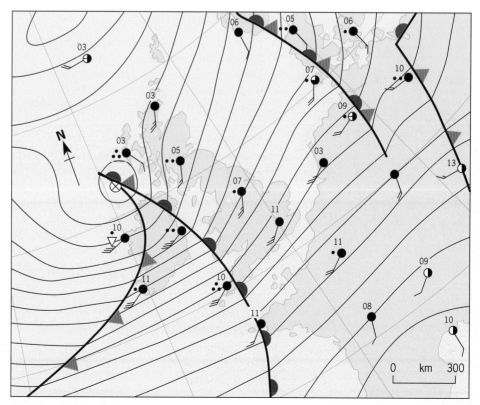

Figure 9.21 Weather map, 2400 hours 25 February 1990 (*Source:* Met. Office)

Towyn floods, Wales

?

11 Using Figures 9.21 and 9.22 describe and suggest reasons for the differences and similarities in the weather between Anglesey and Norfolk.

12 On 25 and 26 February 1990, the Meteorological Office issued warnings of severe storms. How do the station plots and the weather map suggest that this is likely?

13 Assuming the weather pattern moves eastwards at an average speed of 40 km/h, when would you forecast winds from a north-easterly direction that would push water towards the shore in North Wales and encourage flooding?

14 Study Figure 9.19 and suggest how the storm warnings provided by the Meteorological Office might have reduced the impact of the floods in Towyn.

15 What does the following quotation (referring to the storms of early 1990) suggest should be the correct approach to examining the causes and occurrence of such storms? 'Climate shows lots of natural variability and it's extremely difficult to know whether these are a few events at the edge of the normal range or the result of climate changes.' (Professor Bob Hoskins, meteorologist, Reading University)

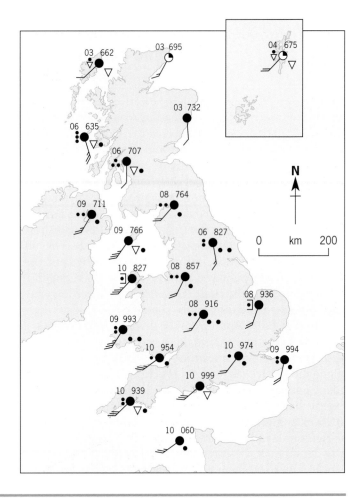

Figure 9.22 Surface observation, 2400 hours 25 February 1990 (*Source*: Met. Office)

9.5 Anticyclones

Anticyclones are areas of *enclosed* high pressure in relation to their surroundings. They usually contain one air mass and so do not involve fronts. As with low pressure systems, anticyclones can occur at different scales although there are basically two types. There are the large and permanent systems that are part of the global pattern of pressure systems. These are often extensions either of the high pressure system associated with the descending air in the tropics, or continental high pressure systems.

The other type is smaller in scale and occurs between depressions. These form, as we have seen, downstream of a **ridge** in the upper westerlies. Convergence at high altitudes results in subsiding air and high pressure with diverging air near the surface. Atmospheric pressure increases towards the centre of an anticyclone and in the northern hemisphere the subsiding air spirals in a clockwise direction. This reflects the influence of the **Coriolis force** (see section 6.2). Sometimes the westerly air streams become so exaggerated that the crests of the ridges break away to form a series of fairly stable anticyclones called **blocking highs** (Fig. 9.23). Blocking highs can slow down and deflect or steer depressions around their edges. The highs are like the areas of calm water in a stream around which whirlpools move.

Characteristics of anticyclones

As the air in an anticyclone is gently descending, it warms adiabatically (see section 7.6), leading to a decrease in relative humidity. The resulting weather depends on the time of year.

In the summer, skies are clear and sunny and heat wave conditions often prevail for several days. Nights may be cold, though, as the lack of cloud cover enables

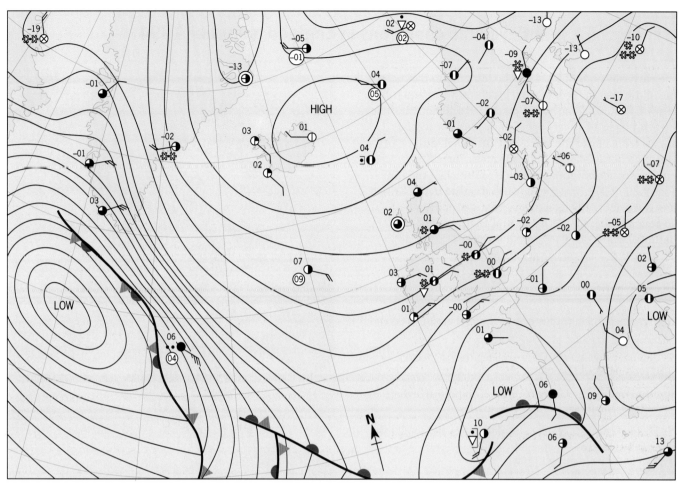

Figure 9.23 Weather map with blocking anticyclones, 7 February 1986

radiant energy to escape. Small quantities of moisture may condense, leading to early morning mist or **radiation fog** (see section 8.3). In the winter anticyclones tend to be more cloudy than in the summer. However, when skies are clear, any **insolation** from the day escapes easily at night rather than being trapped. This leads to very cold temperatures during the day and often several degrees below freezing point at night. It is in these conditions that black ice can form that is such a hazard to road transport (Fig. 9.24).

Figure 9.24 Traffic comes to a halt on the M1 due to heavy snow-fall and the threat of black ice, 8 December 1990

Anticyclonic conditions in England, summer 1994

The early summer of 1994 will be remembered as being one of the best in recent years. June was warm and July, with 197 hours of sunshine, was the second hottest since records began 143 years ago. Conditions throughout August were closer to the average for this month. Although occasional thunderstorms occurred, anticyclonic conditions prevailed for much of the time (Figs 9.25–9.27). Prolonged warm weather in the British Isles does not really lead to hazardous conditions. However, it does affect the natural environment and disrupts a wide range of human activities (Fig. 9.28).

16 Comment on the differences in the type of cloud at A and B in Figure 9.25

17 Identify the isobaric features on Figure 9.26 between A and B and between C and D.

18 How do Figures 9.25, 9.26 and 9.27 indicate that there are anticyclonic conditions over the British Isles?

19 Using Figure 9.27 explain the differences and similarities between the weather being experienced in the south-west of England and East Anglia on 1 July.

20 Study Figure 9.28 and classify the benefits and problems created by the prolonged anticyclonic conditions in the summer of 1994.

Figure 9.25 Weather satellite image, 1 July 1994 (*Source:* Met. Office)

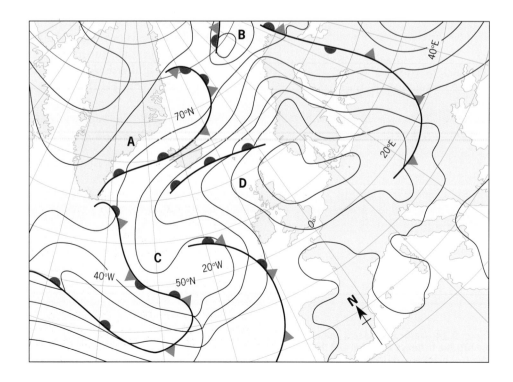

Figure 9.26 Weather map, 0600 hours 1 July 1994 (*Source:* Met. Office)

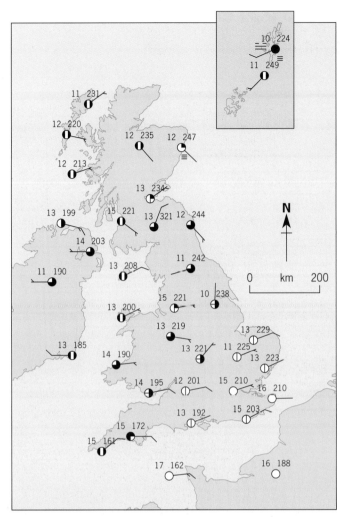

Figure 9.27 Surface observations, 0600 hours 1 July 1994
(*Source:* Met. Office)

Heatwave that produced weather on par with Barcelona and Beirut

LACK of rain resulted in slower growth of grass and so there is less forage available for livestock. Farmers have had to use more expensive food concentrates.

Pigs, meanwhile have been too hot to mate. Farmers were advised to give their herds access to plenty of mud and provide shower units for those kept inside.

The abundance of sunshine has brought the harvest forward by about three weeks because the grain ripened very quickly… Some vineyards are expecting a 50 per cent increase in their crop.

HOT weather boosted sales of soft drinks, beer, snacks, ice creams and fresh produce. Cotton clothes sold particularly well and sales of suntan lotions were higher than last year.

DIY centres reported exceptional demand for gardening equipment, garden furniture and barbecues, although sales of indoor materials suffered.

TRAIN services were disrupted when a section of railway track warped.

An air-conditioning failure trapped passengers on a London to Glasgow train in sweltering temperatures.

There were problems for motorists when tar liquified on roads.

Fierce thunderstorms in mid-August caused traffic chaos in London, with flooded roads and a record 29 Underground stations were closed.

DOCTORS issued renewed warnings about increasing levels of exposure to the sun and warned that tanned skin is damaged skin.

Forty people fainted at a Buckingham Palace garden party and hospitals around the country treated increased numbers of cases of heat exhaustion.

High levels of ozone – which is produced by car exhaust fumes combining with sunlight – left people with asthma and other respiratory problems breathless.

Hospitals reported a 20 or 30-fold increase in admissions for patients with breathing difficulties as the heatwave peaked.

SCORCHING temperatures brought a 4 m sunfish spotted off the Isle of Wight, Potuguese man-o'-war jelly fish and venomous weever fish. In the South West, the warm weather brought basking sharks closer in-shore to popular beaches. A blanket of thunderflies feasted on crops.

Wasps were prolific and pest control officers had to deal with twice as many nests as last year. A population explosion of stoats was blamed for the dearth of Highland grouse. Hedgehogs and badgers died after their food supply of worms and beetles became locked under solid ground.

Figure 9.28 The impact of anticyclonic conditions in July 1994
(*Source:* Kathy Marks, *The Daily Telegraph,* 27 Aug. 1994)

21 Essay: Describe and explain the weather associated with anticyclones and depressions in the winter over the British Isles.

Summary

- The greatest variability in the atmosphere occurs in the middle latitudes.

- Depressions and anticyclones exert a powerful influence on the weather and climate of the mid-latitudes. They can also lead to extreme or hazardous weather conditions that dramatically influence human activities.

- An air mass is a body of air with largely uniform physical properties. A number of different air masses influence the weather in north-west Europe.

- A depression is a low pressure system that brings unsettled weather including strong winds and heavy rain. They particularly influence the weather in the British Isles from late autumn to spring.

- Our understanding of the complex atmospheric phenomena of depressions has improved in recent decades. Research has benefited from the use of radar, satellites and computer models.

- Mid-latitude depressions form an important part of the atmospheric system as they move large quantities of air and energy from low to high latitudes.

- Anticyclones are areas of high pressure that are associated with stable atmospheric conditions. They result in hot, dry weather in the summer and fine, dry but particularly cold weather in the winter.

- A synoptic weather map shows the weather over an area at a given time. A sequence of weather maps used in conjunction with satellite photographs can show the early development, maturity and decay of pressure systems as they pass over an area.

10 Tropical systems

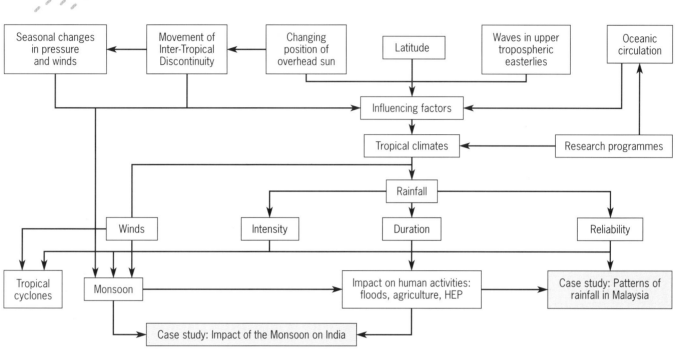

Seasonal changes in pressure and winds ← Movement of Inter-Tropical Discontinuity ← Changing position of overhead sun

Latitude

Waves in upper tropospheric easterlies

Oceanic circulation

Influencing factors

Tropical climates ← Research programmes

Rainfall

Winds | Intensity | Duration | Reliability

Tropical cyclones | Monsoon

Impact on human activities: floods, agriculture, HEP → Case study: Patterns of rainfall in Malaysia

Case study: Impact of the Monsoon on India

Figure 10.1 Parts of Mae Sot, Thailand, become flooded after a tropical rainstorm

10.1 Introduction

Low latitude climates have received much attention from meteorologists since the 1940s. There are now many more weather stations in the tropics and a global weather satellite network has particularly added to our understanding. In the light of new information, meteorologists have modified previous, rather simple, ideas about tropical atmospheric circulation and its uniformity. The World Meteorological Organisation, along with some specialist groups, has carried out a number of research projects on tropical **weather** and **climate**. One such study is called the TOGA Programme (Tropical Ocean and Global Atmosphere, 1985–95). Its purpose is to develop models that predict more accurately the way the tropical atmosphere/ocean system works. It is important to stress the links between the oceans and tropical circulation (see section 6.6), and that the weather and climate of the mid-latitudes is very much influenced by processes in the tropics (see section 9.2). Air movements and ocean currents transfer energy to the higher latitudes.

The mid-latitude climates are characterised by the passage of **depressions** in the westerly air flow. These depressions ensure a fairly even distribution of **precipitation** throughout the year, particularly on the western side of continents. The interiors are drier, but **convectional** storms in the summer provide moisture to offset these otherwise dry conditions. Variations in temperature are more pronounced than fluctuations in rainfall. The climate of the tropics is the opposite, with variations in rainfall not only being a feature of the climate but also having a marked effect on human activities.

The climate of the tropics is also associated with dramatic phenomena like tropical storms, **hurricanes** and the **monsoon.** These have important implications for the people who live in these parts of the world (Fig. 10.1). We will examine the extent to which agricultural practices are related to the extremes of tropical climates. To understand these climatic characteristics it is necessary first to examine the atmosphere in the tropics more closely, and particularly the spatial variations in precipitation.

10.2 General features of tropical climates

The tropics cover a huge area of the globe between 23½° north and south of the equator (see Fig. 9.1). The tropics are defined geographically by the maximum limits of the movement of the overhead (zenithal) sun. From a climatological point of view, the tropics are the area between the semi-permanent sub-tropical high **pressure** cells (see also Figure 9.1). These cells effectively separate the climates of the tropics from those of the mid-latitudes. It is impossible to comment on all the different climatic regimes within the tropics, so we will concentrate on those areas closer to the equator and also the distinctive monsoon climate.

Temperatures and rainfall
While the climate of the mid-latitudes is noted for spatial variations in temperatures, the reverse is true of the tropics. The tropics rarely experience temperatures approaching 0° and **diurnal** changes are often greater than annual temperature ranges. The small range in temperature results from the lack of seasonal variations in the amount of **insolation** – because the sun is always nearly directly overhead (see Fig. 2.5). Rainfall in the mid-latitudes is often prolonged and of low intensity, whereas in the tropics most rainfall occurs during intense storms or showers.

Different **air masses** were also a distinctive feature of the mid-latitudes. In contrast, in the tropics, there is not the diversity of source areas so temperature variations are not very significant in air moving here. Of greater significance are the differences in levels of **humidity** and the vertical motion of the atmosphere. Although we have already studied the general circulation of the tropical atmosphere (see section 6.3), we need to add a few details to understand the tropical systems discussed in this chapter.

The structure of the atmosphere in the tropics
The **tropopause** is higher over the tropics than in the middle latitudes (see Fig. 6.13). Consequently, there may be several different air streams within the space between the earth's surface and the tropopause (Fig. 10.2). For example, one airstream may exist up to 1000 m with a mid-level flow from 1000 to 6000 m; these are easterly movements. There is also a high-level westerly stream above 6000 m which is often strong enough to be referred to as a **jet stream**. These air streams are embedded within a belt of moving air and are often not as clear as the jet streams in the northern hemisphere. They do not mark the boundary between two air masses and therefore use of the term **front** is inappropriate.

Instability and storms

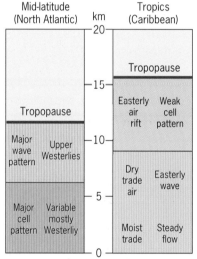

Figure 10.2 Structure of the atmosphere in the mid-latitudes and the tropics (*Source:* Boucher, 1974)

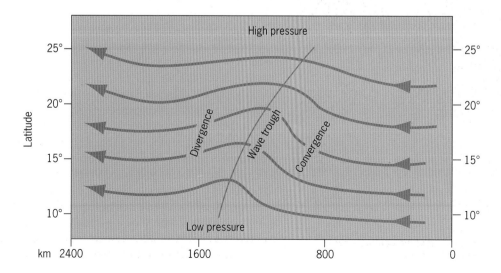

Figure 10.3 Wave in upper atmosphere easterly air currents (*Source:* Boucher, 1974)

1   Explain why annual temperature ranges are small in equatorial and tropical latitudes (use Fig. 2.5 to help you).

2   Suggest why the clouds in Figure 10.5 are associated with heavy rainfall.

Other storms and patterns of weather are associated with wave disturbances occurring in many parts of the tropics, such as the Equatorial Waves in the Pacific and the Easterly Waves in the Caribbean. These appear on surface pressure charts as waves or **troughs** of low pressure in the **isobars** (Fig. 10.3). They have a wavelength of on average about 3000 km and travel from east to west at about 40 km/h, lasting for up to four weeks. As the waves move through the westerly Trades, there is upper air convergence leading to subsiding air ahead of the trough. This leads to fine weather with scattered cumulus clouds. Behind the trough, upper air divergence draws air upwards, causing instability and encouraging cumulonimbus clouds to develop and thundery showers (Figs 10.4 and 10.5). Strong convectional uplift (see section 6.3), resulting from high levels of insolation in the tropics, adds to the **instability**. Although some of these systems can disappear after a few weeks, they may intensify as they move west and so develop into powerful tropical storms or hurricanes. An important point to note is that, in equatorial areas, convectional activity only takes place where unstable conditions encourage air to rise (refer to north-east Brazil in section 13.4).

Figure 10.4 Structure of a storm associated with a Caribbean easterly wave (*Source:* Boucher, 1974)

Figure 10.5 Satellite image of storms over the Caribbean associated with easterly upper troposphere wave

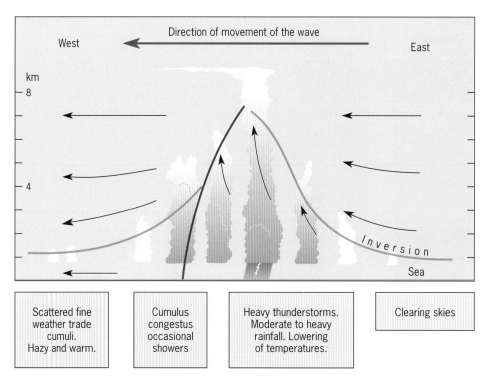

| Scattered fine weather trade cumuli. Hazy and warm. | Cumulus congestus occasional showers | Heavy thunderstorms. Moderate to heavy rainfall. Lowering of temperatures. | Clearing skies |

The Inter-Tropical Discontinuity

The **Inter-Tropical Discontinuity (ITD,** see section 6.3) is another area where air flows converge at a belt of low pressure. This convergence of air often results in strong instability and uplift, leading to towering cumulonimbus clouds and rainfall. Notably, the ITD is not a continuous belt and neither is there an unbroken band of rainfall. The Pacific is the only area where convergence of the trades is strong almost throughout the year. Elsewhere, particularly between July and September, the trades are separated by an area of semi-uniform low pressure called the **doldrums**. This considerably weakens the convergence of the trade winds. The ITD is often quite difficult to locate in certain months, and convergence, cloud cover and rainfall may be weakly developed. Consequently, clear bands of cloud and rainfall do not migrate back and forth between the tropics each year. The ITD is more of a synoptic feature rather than a permanent one. Other factors like relief can also play an important role, e.g. the mountains in Central America and the Andes effectively separate the North-East Trades from the South-West Trades.

Figure 10.6 Landslide after intense rainfall, Colombia

10.3 Patterns of precipitation in the tropics

Rainfall in the tropics is more complex than meteorologists once thought. As we have seen in the previous section, it is influenced by the convergence of the trade winds, changes in the position of the ITD and wave disturbances in the easterly airstreams. There are also less widespread influences like hurricanes, the **El Niño oscillations** (see section 3.4) and the monsoon. In addition, rainfall distribution patterns can reflect local factors, particularly relief and proximity to the sea.

Seasonality of rainfall

The seasonal pattern of rainfall is largely determined by the passage of the ITD. Areas near the equator tend to have a more even distribution of rainfall throughout the year. As we move away from the equator, though, we see the emergence of a clearer wet and dry season (see sections 6.4 and 13.4).

Intensity, duration and reliability of rainfall

In the mid-latitudes rainfall can be prolonged, sometimes lasting intermittently for over 24 hours. However, over the equatorial regions most rainfall comes from showers and intense storms. These are often very localised and short-lived. Subsequently, the concepts of *intensity*, *duration* and *reliability* of rainfall are important in the tropics. Variations of these, in either space or time, can have important consequences for both the natural environment and people. Sudden and intense storms may lead to severe flooding or trigger other natural events like landslides (Fig. 10.6). Although droughts are infrequent, they can occur even in equatorial areas and lead to stress on natural vegetation and crops.

Scale of rainfall

We can also examine rainfall at a variety of scales. We have considered patterns of precipitation at the global and sub-continental scales, in addition to some of the implications for human activities (see section 7.3). Here, we will concentrate on a smaller spatial scale and consider the patterns of rainfall in one tropical country.

Patterns of rainfall in Malaysia

Intensity, duration and reliability of rainfall

Malaysia (Fig. 10.7) is typical of equatorial locations in that rainfall is evenly distributed over much of the country and throughout the year, receiving over 1500 mm (Figs 10.8 and 10.9). There is a slightly wetter period from November to February when strong convergence is taking place over the region. The rising moist air encourages cloud development and rainfall.

In the equatorial trough, or ITD, local relief influences the formation of showers (Fig. 10.10), while storms are also affected by the diurnal pattern of radiation (Figs 10.11–10.12).

Rainfall and agriculture

In Malaysia there is no pronounced dry season and every month experiences high rainfall totals (Fig. 10.13). Surprisingly, despite this, irrigation is used to grow many crops e.g. rice, which is one of the major crops in south-east Asia and forms a staple part of people's diets. Research has shown that at particular times during its

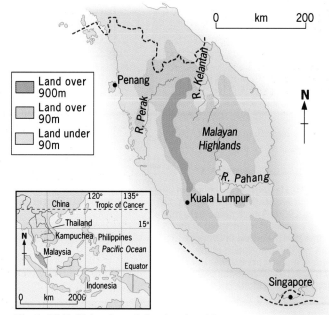

Figure 10.7 Malaysia: relief and major cities

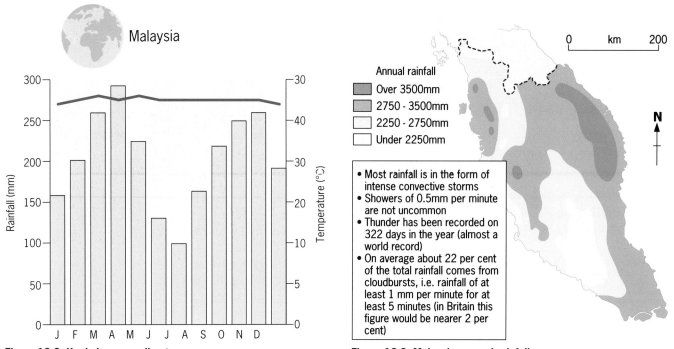

Figure 10.8 Kuala Lumpur: climate

- Most rainfall is in the form of intense convective storms
- Showers of 0.5mm per minute are not uncommon
- Thunder has been recorded on 322 days in the year (almost a world record)
- On average about 22 per cent of the total rainfall comes from cloudbursts, i.e. rainfall of at least 1 mm per minute for at least 5 minutes (in Britain this figure would be nearer 2 per cent)

Annual rainfall
- Over 3500mm
- 2750 – 3500mm
- 2250 – 2750mm
- Under 2250mm

Figure 10.9 Malaysia: annual rainfall

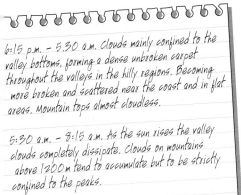

6:15 p.m. – 5.30 a.m. Clouds mainly confined to the valley bottoms, forming a dense unbroken carpet throughout the valleys in the hilly regions. Becoming more broken and scattered near the coast and in flat areas. Mountain tops almost cloudless.

5:30 a.m. – 8:15 a.m. As the sun rises the valley clouds completely dissipate. Clouds on mountains above 1200 m tend to accumulate but to be strictly confined to the peaks.

8:15 a.m. – 11:30 a.m. Continued accumulation of clouds over the steep, narrow ridges above 1000 m, but by 11:30 a.m. beginning to form over narrow ridges down to 500 m.

11:30 a.m. – 4:30 p.m. View obscured by cloud.

4:30–6:15 p.m. Clouds above about 1300 m becoming patchy exposing densely cloud-filled valleys and widespread rain which lessened later in the afternoons. The clouds in the valleys initially very uneven, settling to below about 500 m as an even blanket.

Figure 10.10 Diurnal variations in rainfall in Brunei State (*Source:* Ashton, 1964). These notes were made during a day and night spent at the summit of a mountain in Brunei and are a typical description of the sequence of events leading to rainfall each day. The description could equally apply to Malaysia, as the majority of inland stations show a similar pattern of daily rainfall throughout the year.

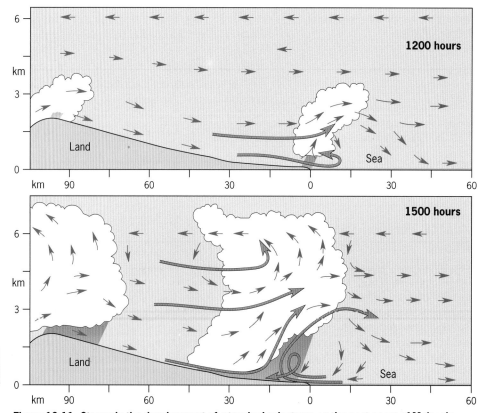

Figure 10.11 Stages in the development of a tropical rainstorm on the east coast of Malaysia

growth, like flowering and the development of the head, rice is very vulnerable to water shortages and yields can suffer. One solution is irrigation, but if people use dirty water, it can lead to toxic substances affecting the crop. Clearly, the availability of a clean supply of water is a priority even in areas experiencing wet tropical climates.

Figure 10.12 Tropical storm clouds

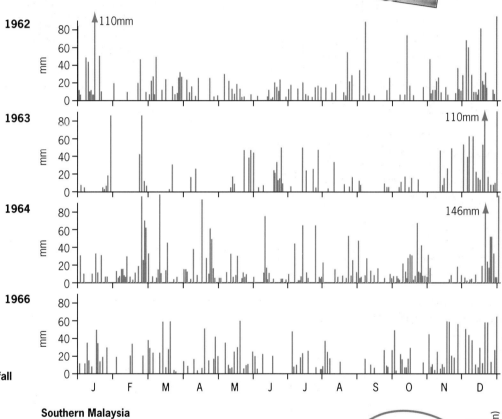

1962

1963

1964

1966

Figure 10.13 Singapore: daily rainfall 1962–66 (*After:* Nieuwolt, 1968)

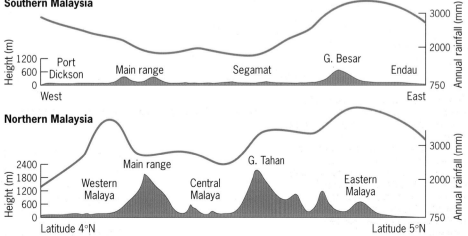

Southern Malaysia

Northern Malaysia

Figure 10.14 The relationship between relief and average annual rainfall in southern and northern Malaysia (*Source:* Lockwood, 1976)

141

Malaysia

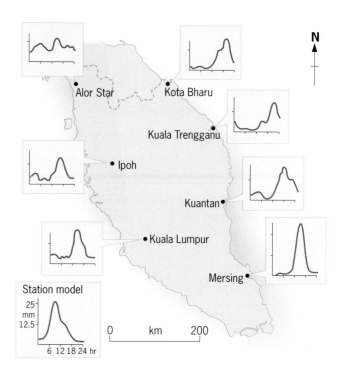

Figure 10.15 Diurnal variations of August rainfall in Malaysia (*Source:* Lockwood, 1976)

_____?_____

3a Trace on to an overlay the pattern of rainfall in Malaysia from Figure 10.9.
b Place your overlay on to Figure 10.7. Describe and explain the pattern of rainfall in Malaysia, referring also to Figure 10.14.

4a Describe the diurnal pattern of rainfall for August shown in Figure 10.15.
b Use the field notes in Figure 10.10 to explain the diurnal pattern of rainfall in Malaysia.

5 Study Figure 10.13.
a What is the minimum amount of rainfall experienced in all months of the year in Singapore?
b Describe the annual pattern of rainfall and also the yearly variations.
c In which year would rice yields have possibly been lowest? Explain your answer.

6 Essay: With reference to tropical wet regions, describe and explain the main features of the climate.

10.4 The monsoon and its causes

The word 'monsoon' comes from the Arabic word for a season. The term once referred to a seasonal reversal of winds in the Arabian Sea, but now includes reversals of pressure and winds and accompanying rainfall. This type of wind system occurs widely over south and south-east Asia, although a similar monsoon system also affects west Africa.

Causes
Firstly, the monsoon has much to do with past geological events. About 15–20 million years ago India collided with the Eurasian plate leading to the uplift of the Himalayas and the Tibetan Plateau (Fig. 10.16). About six million years ago the

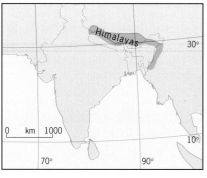

Figure 10.16 Location of the Himalayas and the Tibetan Plateau

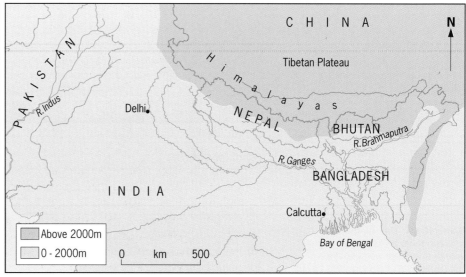

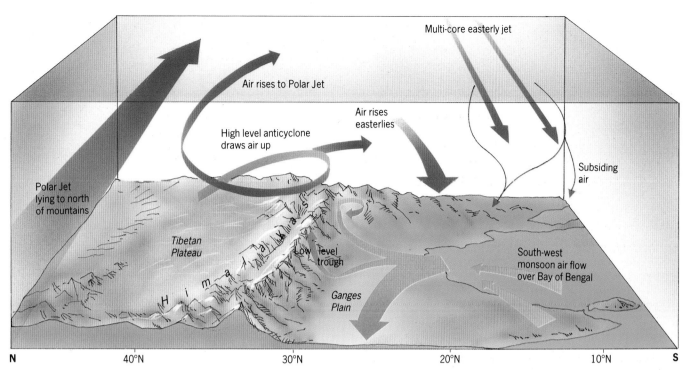

Figure 10.17 Atmospheric circulation features during the south-west summer monsoon, viewed eastwards across central India

Himalayas reached such a height that they affected the atmospheric circulation and set up new patterns of movement over the Indian sub-continent. In the 1980s, scientists using computer simulations confirmed the importance of the Himalayas. Without the mountains' influence, the movement of the ITD would be different and so patterns of rainfall over much of India and south Asia would result in more desert-like conditions.

In addition, scientists have argued that the plateau was responsible for reducing the concentration of carbon dioxide (CO_2) in the atmosphere. CO_2 absorbs heat radiated from the earth's surface, so as CO_2 levels dropped they argue that less radiated energy from the earth was trapped and the climate began to cool. The mechanism suggested by Maureen Raymo involves several influences:

- The great height of the plateau influences patterns of air circulation that draw warm, moist air from the Indian Ocean (Fig. 10.17). This produces the Indian monsoon rains in the summer that also affect the southern edge of the plateau.

- CO_2 dissolves in the rain. As more uplift occurs and more warm air is drawn from the Indian Ocean, more rainfall results and more CO_2 is taken out of the atmosphere. Chemical weathering by the acidic rain and erosion of the plateau's rocks return the CO_2, in the form of calcium bicarbonate, to the oceans.

Other scientists have made alternative suggestions:
- The removal of CO_2 from the atmosphere might in part have been balanced by slower rates of weathering in other parts of the world due to a cooler climate contributing to a new stable cycle of carbon.

These ideas that help to explain the occurrence of the monsoon are only expressed relatively simply here. Although many scientists broadly accept Raymo's work, this is an area that requires further study.

Secondly, the monsoon also results from the differential heating and cooling of land in comparison with adjacent sea areas. This in turn results in different pressure and wind systems (Fig. 10.18). The heating effect is about six times greater over the land because over the sea most energy is used in **evaporation**. Thirdly, the

143

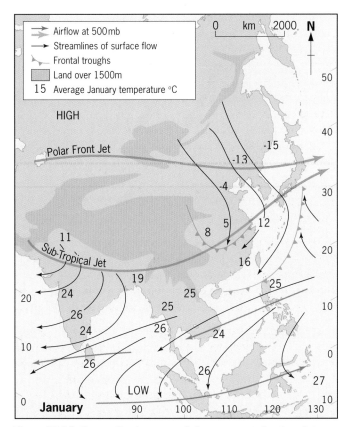

Figure 10.18 Generalised pattern of air movements for the winter north-east monsoon

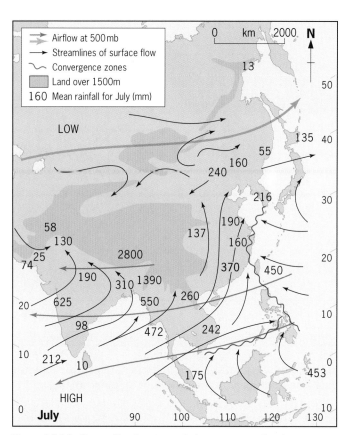

Figure 10.19 Generalised pattern of air movements for the summer south-west monsoon

monsoon is in part due to the movement of the ITD and its wind belts during summer in the northern hemisphere.

The characteristics of the monsoon climate are particularly clear in India and Bangladesh. Here the seasonal changes can be broadly divided into two periods: the *north-east monsoon* of the winter and spring and the *south-west monsoon* of the summer and autumn.

The winter: north-east monsoon

During the northern winter the westerly jet stream splits either side of the Tibetan Plateau (Fig. 10.18). The southern part of this jet is affected in its position by the Himalayas to the north. Air descends from this air stream giving rise to an extensive area of high pressure near the surface over much of central Asia. This leads to outward-blowing, north-easterly winds that cross south Asia. These dry airstreams produce clear skies and sunny weather over most of India from November to May. Little rain falls at this time of year although slightly higher amounts of rainfall in south-east India and eastern Sri Lanka result from moisture evaporated from the Bay of Bengal (this area is crossed by the north-easterly air currents). At this time, the ITD has moved south of the equator with the overhead sun.

The summer: south-west monsoon

The so-called 'burst' or start of the monsoon is linked to a number of factors operating in the **troposphere** that begin in the spring and culminate around early June.

1 The first significant change occurs between March and May: the upper westerly air currents begin to move north, the northerly section of the jet stream strengthens and the southerly part weakens until eventually the jet stream lies wholly to the north of the Himalayas (Fig. 10.19).

2 As the overhead sun migrates over India, the ITD moves to a position just south of the Himalayas. This is also referred to as the monsoon trough – a zone of low pressure that progresses northwards in a series of pulses. There are two clear sections to the monsoon trough as it progresses northwards at different speeds (Fig. 10.19). One passes over the Bay of Bengal, the other further west over India. The monsoon breaks first over south-west India in the later half of May. The east coast of India and Bangladesh receive the monsoon towards the end of the month. The two sections of the trough meet over north central India in mid-June before moving on towards the north-west of the country and Kashmir.

3 Strong convectional heating under clear skies also contributes to the development of a large area of surface low pressure. Warm, moist air is drawn in from over the Indian Ocean and marks the start of the monsoon rains. This is one of the major forces in the development of the monsoon.

4 The low pressure encourages rising air over northern India, Bangladesh and the Himalayas. This leads to an upper tropospheric **anticyclone** (high pressure) with outwards-moving air. The **Coriolis force** (see section 6.2) influences these air currents and a strong westerly airstream develops in the lower troposphere. In the upper troposphere a strong easterly jet stream expands rapidly over southern Asia.

These influences combine to draw in air from over the Indian Ocean and Arabian Sea that then rises to the upper troposphere where it passes out of the region. Much of the moisture involved is picked up as a result of evaporation over the ocean areas. The warm, moist air is encouraged to rise either by convection or **orographic** influences to form clouds and rainfall. A great deal of **latent heat** is released and this further encourages the convectional processes and reinforces the monsoon.

Incidentally, this section has further brought into question the appropriatness of the term Inter-Tropical Discontinuity. You may have realised that the ITD at certain times of the year bends over parts of southern Asia outside of the tropics!

?

7 Construct a process/flow diagram to summarise the main stages in the development of:
a the winter north-east monsoon
b the summer south-west monsoon.

10.5 The nature of the monsoon

The climate data for Calcutta (Table 10.1) clearly reflects the influence of the monsoon in the main area where it occurs. The arrival of the summer monsoon is not as sudden, though, as some writers suggest (Fig. 10.20).

Table 10.1 Calcutta, India: climate statistics

Months	Jan	Feb	Mar	April	May	June	July	Aug	Sept	Oct	Nov	Dec
Temperature average (°C)	20.2	23.0	27.9	30.1	31.1	30.4	29.1	29.1	29.9	27.9	24.0	20.6
Precipitation (mm)	13	24	27	43	121	259	301	306	290	160	35	3

'You have never seen the monsoon's burst? In Bombay it is quite something. For months the city has been absolutely sweltering and then, usually on the afternoon of 10 June, huge clouds begin to build up over the sea. Soon your wind comes, so strong that it will sink any little boats that haven't taken shelter. The wind drops, it gets very dark, there is terrific thunder and lightning and then – the deluge! Suddenly the air is very cool and perfumed with flowers. It is time of rejoicing. And renewal.' She frowned at me. 'It is also when I feel perhaps most truly Indian.

Figure 10.20 The arrival of the monsoon (*Source: Frater A, Chasing the Monsoon*, 1991, Penguin Books)

As the monsoon moves northwards, this movement is not a smooth progression but rather a stepped one with the monsoon pausing, or 'stagnating' for a while, as it advances over India (Fig. 10.22). Although the monsoon may begin abruptly, in some locations drizzle often precedes the onset of heavy rain and monsoon winds may arrive with no accompanying rainfall. Not only can the arrival be unspectacular but also unreliable; data published by the Indian Meteorological Department have shown a standard deviation of between five and ten days. Occasionally, as in 1987, the monsoon rains fail in some parts of India and severe droughts occur while other regions experience floods. The south-west monsoon begins to withdraw in September leading to the onset of the north-east monsoon (Fig. 10.23).

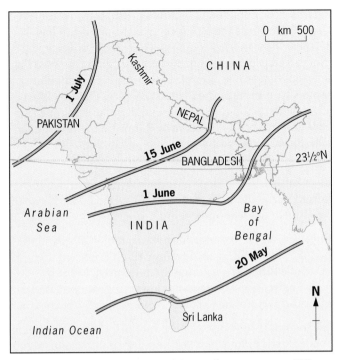

Figure 10.21 Dates of the onset of the south-west monsoon (*After:* Subbaramayya et al., 1984)

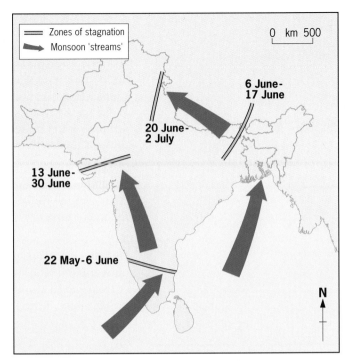

Figure 10.22 The stepped progress of the south-west monsoon over the Indian sub-continent (*After:* Subbaramayya et al., 1984)

?
---

8 Use Figure 10.21 to describe when the wet season in south Asia begins.

The monsoon is not simply a prolonged period of exceptional rainfall, for there are also drier periods lasting several weeks, mainly in June and August. In addition, there are several other atmospheric disturbances associated with the monsoon. For example, two or three large monsoon depressions, 200–500 km in diameter, usually occur each month. They last for several days and account for much of the monsoon rainfall. Also, subtropical **cyclones** form between mid-June and mid-September over the Arabian Sea. These are larger, 300–600 km in diameter, and are responsible for much of western India's rainfall at this time.

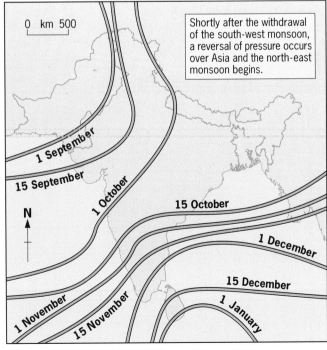

Figure 10.23 Dates of the withdrawal of the south-west monsoon

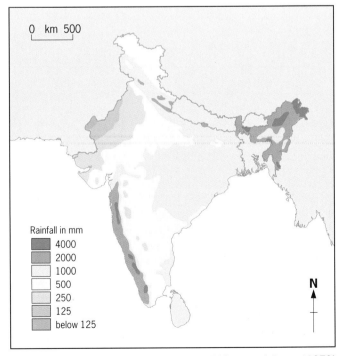

Figure 10.24 India: average annual rainfall (*Source:* Johnson, 1979)

Figure 10.25 India: relief and main cities

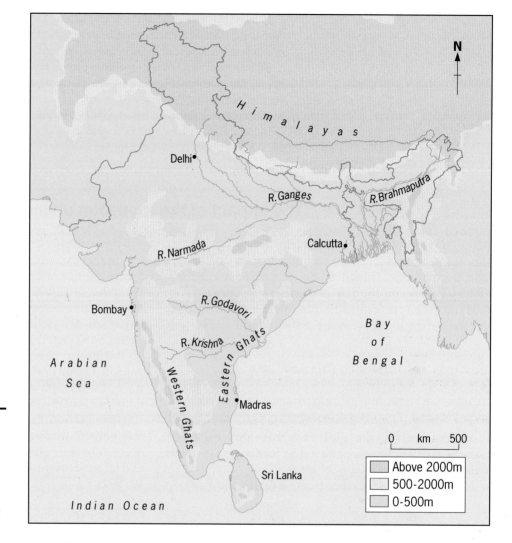

?

9 Use Figures 10.24 and 10.25 to describe and explain the pattern of rainfall in India.

10a Draw a climate graph based on the data in Table 10.1
b How does your graph reflect the influence of the monsoon?
c Account for the fall in average monthly temperature in July and August in comparison with April, May, June and September in Calcutta.

11a Use Table 10.2 to plot a line graph based on the data for Bombay and New Delhi.
b On your completed line graph, plot two curves for the three-year running means for Bombay and New Delhi.
c Calculate the standard deviations for Bombay and New Delhi.
d Use your graph and calculations to comment on the variability of the rainfall and the monsoon in Bombay and New Delhi. Refer to years in which flooding or drought conditions might have been particularly severe. Remember that not all of India is affected equally in a given year by either drought or floods.

Table 10.2 Bombay and New Delhi, India: rainfall (*Source*: CRU, University of East Anglia)

Year	New Delhi: annual rainfall	Bombay: annual rainfall	Year	New Delhi: annual rainfall	Bombay: annual rainfall
1950	871	1562	1973	667	1850
1951	444	1788	1974	542	2470
1952	520	1608	1975	1201	2792
1953	689	2376	1976	995	1882
1954	709	3481	1977	926	2351
1955	804	2251	1978	1040	1835
1956	853	2572	1979	433	1832
1957	809	1775	1980	686	1923
1958	1034	3319	1981	613	2440
1959	555	2393	1982	857	2381
1960	764	2118	1983	1059	3107
1961	1108	2206	1984	695	2140
1962	579	2029	1985	nd	2423
1963	802	2530	1986	nd	1287
1964	1232	1964	1987	380	1937
1965	593	2024	1988	nd	nd
1966	652	1560	1989	429	1900
1967	1069	2373	1990	1050	nd
1968	612	960	1991	nd	2620
1969	682	1552	1992	nd	1629
1970	748	2625	1993	985	2370
1971	907	2443	1994	nd	nd
1972	864	1511			

nd = no data available   Figures in mm

Hurricanes also develop over the Indian Ocean and the Bay of Bengal between April and June and late September to early December. Other smaller low pressure systems can also occur, producing further rainfall.

The rainfall pattern is not simple either, with considerable spatial variability across the country (Fig. 10.24). The Ganges valley experiences two or three westward moving depressions per month in the summer. Without these, rainfall totals would be reduced because the low-lying relief does not encourage uplift. Understanding the monsoon is vital to aid forecasting and better management of such consequences as flooding.

The impact of the monsoon on India

Between May and October much of the Indian subcontinent receives 80 per cent of its annual rainfall from the monsoon. This affects the lives of about a quarter of the world's population in southern Asia and has dramatic social and economic consequences for countries like India (Figs 10.25, 10.26, 10.28, 10.31 and 10.32).

Agriculture

The total amount, seasonal incidence and the spatial distribution of rainfall (Figs 10.24 and 10.27) are the most significant factors influencing agriculture in India (Fig. 10.28). A huge range of crops is grown in India but we will concentrate on the main subsistence crops (grown to feed farmers and their families rather than for sale or export) to illustrate the importance of the monsoon.

Throughout the 1960s the monsoon rains were poor

and the reduced rainfall affected much of north-east India. The worst years were in the mid–1960s (Fig. 10.30). Severe drought affected several states including some like Bihar that usually have over 1000 mm of rain a year. This dramatically affected crop yields. For example, in Rajasthan, rice yields fell by as much as 72 per cent and in Orissa by 81 per cent. Yields of other crops fell by between 20 and 30 per cent in most of the affected states. The general picture in Figure 10.30 hides local situations where the drought was even more serious, such as in the rainshadow of the Western Ghats.

In comparison, the early 1970s were years when rainfall was in many areas significantly above normal. Yields per hectare for rice, wheat and bajra in Karnataka were 7, 12 and 62 per cent respectively above the average for 1970–2.

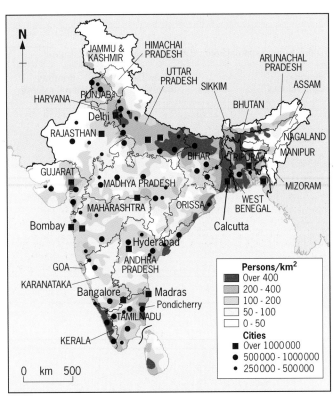

Figure 10.26 India: political and population distribution

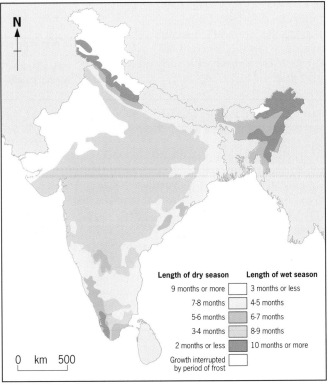

Figure 10.27 India's wet and dry seasons

?

12 Describe and account for the changes in crops and farming practices from west to east along the transect in Figure 10.29. Use Figures 10.24 and 10.27 to help you.

13 To what extent do you think the monsoon affects agriculture in India? You could use a reference book to find the ideal conditions for the growth of rice and cotton.

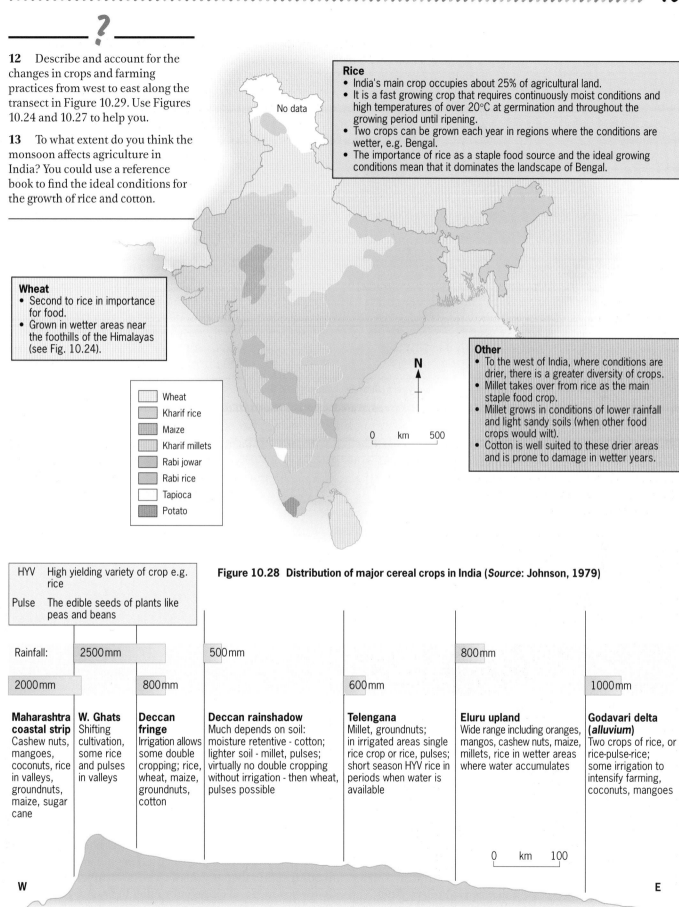

Rice
- India's main crop occupies about 25% of agricultural land.
- It is a fast growing crop that requires continuously moist conditions and high temperatures of over 20°C at germination and throughout the growing period until ripening.
- Two crops can be grown each year in regions where the conditions are wetter, e.g. Bengal.
- The importance of rice as a staple food source and the ideal growing conditions mean that it dominates the landscape of Bengal.

No data

Wheat
- Second to rice in importance for food.
- Grown in wetter areas near the foothills of the Himalayas (see Fig. 10.24).

N

Other
- To the west of India, where conditions are drier, there is a greater diversity of crops.
- Millet takes over from rice as the main staple food crop.
- Millet grows in conditions of lower rainfall and light sandy soils (when other food crops would wilt).
- Cotton is well suited to these drier areas and is prone to damage in wetter years.

Wheat
Kharif rice
Maize
Kharif millets
Rabi jowar
Rabi rice
Tapioca
Potato

0   km   500

| HYV | High yielding variety of crop e.g. rice |
| Pulse | The edible seeds of plants like peas and beans |

Figure 10.28 Distribution of major cereal crops in India (*Source*: Johnson, 1979)

Rainfall: 2500mm   500mm   800mm

2000mm   800mm   600mm   1000mm

Maharashtra coastal strip
Cashew nuts, mangoes, coconuts, rice in valleys, groundnuts, maize, sugar cane

W. Ghats
Shifting cultivation, some rice and pulses in valleys

Deccan fringe
Irrigation allows some double cropping; rice, wheat, maize, groundnuts, cotton

Deccan rainshadow
Much depends on soil: moisture retentive - cotton; lighter soil - millet, pulses; virtually no double cropping without irrigation - then wheat, pulses possible

Telengana
Millet, groundnuts; in irrigated areas single rice crop or rice, pulses; short season HYV rice in periods when water is available

Eluru upland
Wide range including oranges, mangos, cashew nuts, maize, millets, rice in wetter areas where water accumulates

Godavari delta (*alluvium*)
Two crops of rice, or rice-pulse-rice; some irrigation to intensify farming, coconuts, mangoes

0   km   100

W   E

Figure 10.29 India: west-east coast agriculture

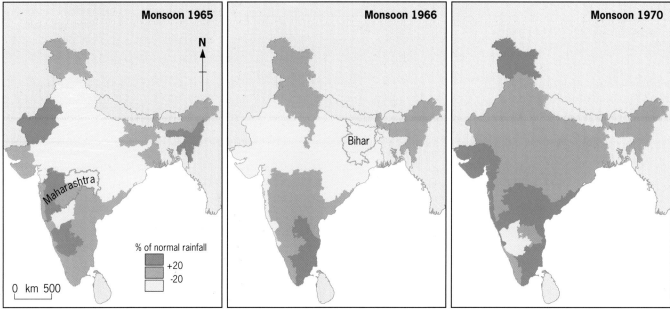

Figure 10.30 Variations in monsoon rainfall for selected years (*Source*: Johnson, 1979)

?

14a Complete a line graph for the data in Table 10.3

b Describe and account for the changes in production of the three cereal crops.

c Use your graph to comment on the likely effects that changes in total grain production may have had on the national economy.

Table 10.3 India: cereal production and imports (*Source*: Johnson, 1979)

Year	Rice	Wheat	Bajra (millet)	Total (all cereals)	Imports
1950–1951	22.1	6.8	2.7	45.8	2.1
1960–1961	34.6	11.0	3.3	69.3	3.8
1961–1962	35.7	12.1	3.6	71.0	2.4
1962–1963	33.2	10.8	4.0	68.6	3.0
1963–1964	37.0	9.9	3.9	70.6	3.3
1964–1965	39.3	12.3	4.5	76.9	6.9
1965–1966	30.6	10.4	3.8	62.4	7.7
1966–1967	30.4	11.4	4.5	65.9	10.4
1967–1968	37.6	16.5	5.2	83.0	8.7
1968–1969	39.6	18.7	3.8	83.6	5.7
1969–1970	40.4	20.1	5.3	87.8	3.9
1970–1971	42.2	23.8	8.0	96.6	3.6
1971–1972	43.1	26.4	5.3	94.1	2.1
1972–1973	38.6	24.9	3.8	87.1	0.4
1973–1974	44.0	30.0	6.5	93.9	3.6
1974–1975*	40.3	24.2	3.2	90.7	4.9

*estimate   Figures in million tonnes

Hazards created by the monsoon

Flooding in India is an annual hazard affecting nearly all the states, but particularly those that experience the heaviest monsoon rainfall (see Fig. 10.24). Of all the regions in India, it is the Ganges floodplain that experiences the worst flooding (Fig. 10.31). This is a monotonous region being physically almost flat from the Bay of Bengal to the foothills of the Himalayas. It is a huge alluvial plain that is intensively farmed and is also one of the most densely populated areas of the world.

Every year there is extensive damage to property and crops as well as loss of life (Fig. 10.32). In 1993 the heaviest monsoon for several years occurred. The death toll exceeeded 650 people and affected 12 states including much of New Delhi, the capital city. Road and rail links were cut and hundreds of thousands of hectares of farmland were under water.

The Indian government has introduced some flood control measures such as embankments along the edge of the Ganges. However, India is an economically developing country. The lack of resources combined with the scale of the hazard, have caused people to adjust to the flooding rather than prevent it. The government gives flood warnings and emergency relief but many people often protest against the government because of its feeble rescue efforts.

Indian army joins rescue as 350 die in monsoon

The Indian army, airforce and navy joined large-scale rescue and relief operations in northern India yesterday after heavy monsoon rains caused floods and landslides, killing nearly 350 people.

According to official figures, at least 323 people have died in the states of Kashmir, Himachal Pradesh, Punjab and Harayana. More are feared dead and hundreds are missing. The rains have been incessant for at least 10 days, causing landslides… Debris has blocked rivers, seriously damaging at least three power stations. Many areas have been blacked out. In Punjab, parts of the cities of Jullundur and Amristar are waist-deep in water.

Figure 10.31 Housing flooded by the Ganges

Figure 10.32 Hazardous conditions created by the monsoon in 1993 (*Source: The Daily Telegraph*, 17 July 1993)

Table 10.4 Adjustments to floods in Bajura Narang, India (*After:* Ramachandran and Tahkur, 1974)

Adjustments	Mentioned by the respondent	Yes, when asked	No, when asked
1 Evacuation by foot	47	3	16
2 Reach for roof top/treetop	39	16	11
3 Storing food grains above ground	15	46	5
4 Go to a) temple b) dormitory c) relative's place d) others – outside floodplain	14	11	41
5 Raised floors	4	42	20
6 Build brick houses with flat roofs for refuge	5	36	25
7 Wait for government evacuation	3	10	53
8 Evacuation before start of floods (of household items, cattle, children, old people etc.)	1	2	63
9 Go to own house outside floodplains	1	2	63
10 Planting trees near house in fields	1	15	50
11 Prior arrangement with private boatmen	0	12	54
12 Use private/neighbour's boat	0	0	66

15 The monsoon itself is not particularly hazardous. Study Figure 10.32 and identify the hazards that are created by heavy monsoon rainfall.

16 Table 10.4 contains the results of a survey conducted in a village in the upper Ganges valley. The population is generally poor and the level of literacy is low. How do the answers reflect the interviewees perception of the flood hazard?

17 With reference to India or Bangladesh, outline the influence of the monsoon with specific reference to:
a social impacts.
b economic impacts.

Summary

- Spatial and temporal variations in precipitation are more important features of tropical climates than fluctuations of temperature.

- In the tropics, most rainfall occurs as showers or intense storms rather than falling over long periods. This is largely the result of convectional processes.

- The Inter-Tropical Discontinuity is a discontinuous zone where the trade winds converge. It is a region of strong upward-moving air currents and frequently unstable conditions leading to tropical storms.

- Waves in the upper tropospheric easterlies can intensify to form powerful storms including hurricanes.

- The seasonal pattern of rainfall is largely determined by the passage of the ITD. Areas near the equator tend to have a more even distribution of rainfall throughout the year. With increasing distance from the equator, a more pronounced wet and dry season emerges.

- Over the equatorial regions, most rainfall comes from showers and intense storms. They are often very localised and short-lived. The concepts of intensity, duration and reliability of rainfall are important in the tropics. Variations in these, either spatially or in time, can have important consequences for the natural environment and people.

- The monsoon is a seasonal reversal of pressure and winds and accompanying rainfall. This type of wind system occurs widely over south and south-east Asia, although a similar monsoon system also affects west Africa.

- The monsoon has dramatic social and economic consequences for the people living in much of south-east Asia. This includes loss of life as a result of severe flooding.

11 Hurricanes – a major hazard

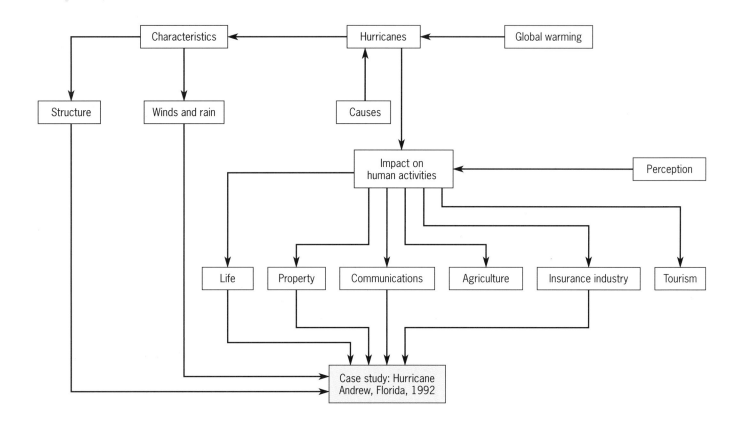

11.1 Introduction

Hurricanes have caused more deaths worldwide than any other form of natural disaster (except perhaps droughts). The effects of a hurricane can be catastrophic, causing widespread damage to property, disrupting communications and causing loss of life (Fig. 11.1).

Figure 11.1 Pleasure-craft tossed out of the water by Hurricane Hugo, South Carolina, USA, 22 September 1989

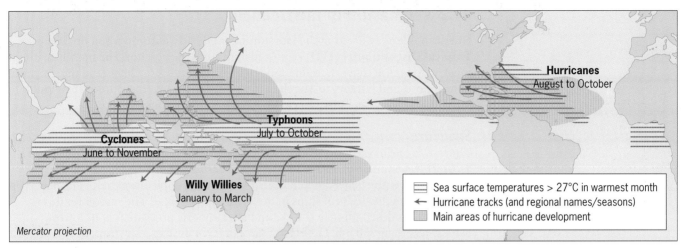

Figure 11.2 The distribution of hurricanes or tropical cyclones

People in different parts of the world give these atmospheric disturbances various names (Fig. 11.2), although we should strictly call them tropical **cyclones**. Tropical cyclones can be classified as follows:

1 *Tropical **depression**:* an organised system of clouds and thunderstorms with a defined circulation and maximum sustained winds of 61 km/h or less.
2 *Tropical storm:* an organised system of strong thunderstorms with a defined circulation and maximum sustained winds of 61–118 km/h.
3 *Hurricane:* an intense tropical **weather** system with a well defined circulation and maximum sustained winds of 119 km/h or higher.

We will use the term hurricane as this is the term used in Europe. Hurricanes are intense low **pressure** systems largely restricted to ocean areas 30° north and south of the equator. They travel westwards before turning away from the equator (Fig. 11.2). The path taken by a hurricane is called a track. From space, astronauts see hurricanes as spiral clusters of cloud up to 800 km across (Fig. 11.3). We will examine the characteristics of hurricanes and how they develop. Finally, we will consider how some scientists have argued that a number of exceptionally powerful hurricanes in recent years may be a result of **global warming**.

We will also consider the impact that hurricanes have on people's lives. One of the most devastating long-term impacts has been on the insurance industry. We will examine the extent to which the changing fortunes of this industry can be attributed to hurricane damage. Hurricane Andrew, which devastated south Florida in 1992, will be used as a case study. This was a Category 5 storm (Table 11.1) which is the maximum rating for hurricanes. Andrew was one of the fiercest hurricanes to hit the USA in the twentieth century.

Figure 11.3 The eye of Hurricane Elena, photographed from space shuttle *Discovery*, 2 September 1985

Table 11.1 The Saffir-Simpson hurricane scale (*Source*: US Dept of Commerce)

Scale number (category)	Sustained winds (km/h)	Damage	Examples, with states affected
1	79–102	Minimal	Florence (1988), Louisiana Charley (1988), North Carolina
2	103–118	Moderate	Kate (1985), Florida Bob (1991), Florida
3	119–139	Extensive	Alicia (1983), Texas Emily (1993), North Carolina
4	140–166	Extreme	Hugo (1989), South Carolina Andrew (1992) Florida, Louisiana, Texas
5	>166	Catastrophic	Labor Day Hurricane (1935), Florida Keys Camille (1969), Louisiana, Mississippi

11.2 The nature of hurricanes

Hurricanes largely occur in the late summer and early autumn when the **Inter-Tropical Discontinuity** (**ITD**) is most distant from the equator. The months with most hurricane activity therefore tend to be from August to October in the northern hemisphere, and January to March in the southern. On average, there are about 80 hurricanes each year, mostly occurring in the northern hemisphere.

Structure

The '*eye*' is one of the most distinctive features of the structure of a hurricane (Fig. 11.4). The eye can be between 20 and 60 km across where there is a calm and warmer area with clear skies and little cloud cover. People have even been observed sunbathing on the decks of ships in the eye of a hurricane! The reason for this is because air is descending in the eye and atmospheric pressure is at its lowest, usually less than 950 mb. *Centrifugal force* (a force pulling away from the centre of a revolving object) associated with the rapid rotation helps to maintain the eye while the spinning pulls clouds and air away from the centre. As the intensity of a hurricane increases and the low pressure intensifies, the eye decreases in size.

Around the eye is an area of towering cumulonimbus clouds stretching up to 15 km into the atmosphere. These clouds form an unbroken wall at the edge of the eye and become more fragmented and lower towards the outer edge of the hurricane. This area of cloud is composed of up to 200 separate convection cells (see section 6.3). Intense **instability** causes rapid rising of air and encourages cloud development. As air rises, it cools and the water vapour in the air **condenses**. Energy is released as the moisture changes from a gas to a liquid, known as **latent heat**, and it is this energy which helps to drive the hurricane.

Winds and rain

Hurricanes are also associated with intense wind speeds which must average at over 119 km/h for a storm to be classified as a hurricane (see Table 11.1). Wind speeds may be much higher, though, with gusts reaching over 300 km/h. The exceptionally low atmospheric pressure and steep **pressure gradients** are responsible for the strong winds. Hurricanes can also produce tornadoes that can add to their

---

?

1 Use the satellite image in Figure 11.3 to draw an annotated diagram outlining the main features of a hurricane.

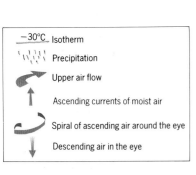

−30°C Isotherm

Precipitation

Upper air flow

Ascending currents of moist air

Spiral of ascending air around the eye

Descending air in the eye

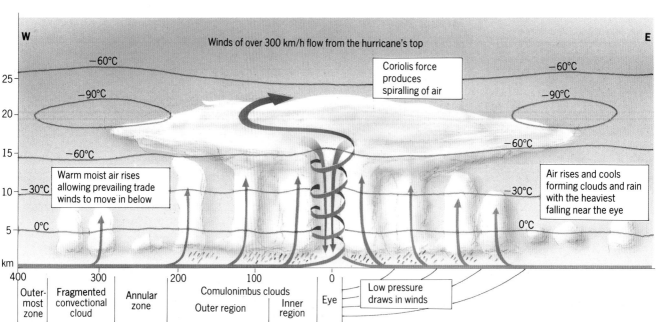

Figure 11.4 The development and structure of a hurricane

destructive power. Tornadoes usually occur in thunderstorms embedded in bands of rain away from the centre of a hurricane, although some do develop near the eye wall.

Heavy rainfall also occurs, typically between 10 and 25 cm in a day, although in extreme cases 100 cm has been recorded. In some tropical countries hurricanes make an important contribution to total annual **precipitation**.

11.3 The formation of hurricanes

Figure 11.2 shows that hurricanes develop over specific parts of the tropical oceans. In fact, major tropical storms often begin as small areas of low pressure along the ITD. Here, air from the northern and southern hemispheres meets and there is strong uplift of air currents. In a year, only four or five out of about forty seedlings (hurricanes in the early stage of development) in the Atlantic may develop into full hurricanes. There are a number of specific conditions, though, necessary for the formation of hurricanes that explain why not all tropical storms turn into hurricanes.

1 Suitable atmospheric conditions are needed, including a large area of more or less uniform temperature, **humidity**, and pressure. Conditions in the upper **troposphere** need to be such that air drawn into the storm near the surface can flow out at a higher altitude. Wind velocities at different altitudes also need to be similar, otherwise the embryonic hurricanes are pulled apart. Meteorologists believe that hurricanes originate from waves in the tropical easterlies or trade winds. During the hurricane season two easterly waves drift over the Atlantic each week.

2 Hurricanes form over warm seas. A large ocean area is therefore needed with average surface temperatures greater than 27°C and a warm water layer at least 60 m deep.

3 A location at least 5° north and south of the equator is important. Here the **Coriolis force** (see section 6.2) is sufficient to enable considerable rotation to occur.

4 The sub-tropical **anticyclones** tend to have strong subsidence on their eastern sides and more suitable conditions for **convection** on their western edges. Thus hurricanes tend to develop on the western sides of oceans.

5 **Relative humidity** needs to be greater than 60 per cent in order for enough energy to be released to drive the hurricane. This energy is produced when water vapour is converted back to a liquid.

6 Air pressures and winds speed need initially to be low at ground level but high in the upper atmosphere for the hurricane vortex to develop.

Research has shown that hurricanes begin as a disturbance in the easterly waves. Air is drawn into the low pressure system that then rises and cools (see Fig. 11.4). We have seen previously that cold air holds less moisture than warm air (see section 7.5) and that as condensation occurs latent heat is released. Huge amounts of latent heat concentrated in a small area can have the effect of triggering a further fall in air pressure at the surface. The low pressure then intensifies and in turn draws in more moist air from the warm ocean. This again rises, leading to condensation and the release of more latent heat.

The rotation of the earth and the Coriolis force then cause the rising air to spin and so create a small compact tropical storm. The distinctive spiral pattern seems to be the result of swirling storm bands, although satellite observations suggest that the individual storm, or convection, cells (see section 11.2) move outwards along the spiral band and slowly decay. Over a period of several days these processes will lead to the development of a hurricane.

?

2a Use Figure 11.2 and Table 11.2 to describe and explain the tropical areas of origin for hurricanes.
b Use Figure 2.27 to suggest why there are no hurricanes off the west coast of South America or in the South Atlantic.

3 Hurricanes weaken if they:
• cross cooler areas of ocean,
• move over an area of land and
• draw in cooler air from a different area. Explain why this should happen.

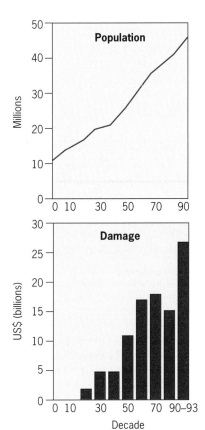

Figure 11.6 Population growth and damage from hurricanes (coastal county populations, Texas to Maine, 1990) (*Source:* NOAA)

Table 11.2 The average frequency of tropical storms of over force 8 by region

Ocean	Jan	Feb	Mar	Apr	May	June	July	Aug	Sept	Oct	Nov	Dec	Annual average
N Atlantic	–	–	–	–	–	0.3	0.4	1.5	2.7	1.3	0.3	–	5.2
NE Pacific	–	–	–	–	0.3	0.6	0.9	2.0	1.8	1.0	–	–	5.8
NW Pacific	0.3	0.2	0.2	0.7	0.9	1.2	2.7	4.0	4.1	3.3	2.1	0.7	17.8
SW Pacific	0.7	1.1	1.3	0.3	–	–	0.1	0.1	–	–	0.3	0.5	3.8
SW Indian	1.3	1.1	0.8	0.4	–	–	–	–	–	–	–	0.5	3.8
N Indian	–	–	–	0.1	0.5	0.2	0.1	–	0.1	0.4	0.6	0.2	2.2

11.4 The impact of hurricanes

The power of a hurricane is enormous – a large hurricane releases more energy in a day than the USA consumes in a year. Throughout history, hurricanes have taken an enormous toll: in 1737 a wall of water killed about 300 000 people in Calcutta and in November 1970 a cyclone struck the Bay of Bengal and swept as many as 400 000 people to their deaths. According to the World Meteorological Organisation, about 50 countries with a total population of 500 million are exposed to occasional strikes by hurricanes. In an average year, about 20 000 deaths occur worldwide as a result of hurricanes.

Although the powerful winds and intense rainfall cause a huge amount of damage, by far the most dangerous feature of a hurricane is the '**storm surge**' (Fig. 11.5). This occurs when the eye of the hurricane, with its attendant very low pressure, raises the surface of the sea by up to 10 m. This can produce exceptionally high seas and waves. If this occurs with a high tide and reaches a low-lying coast, (the worst possible combination), vast areas of land may be flooded. Historically, storm surges have been the feature of hurricanes responsible for the greatest loss of life.

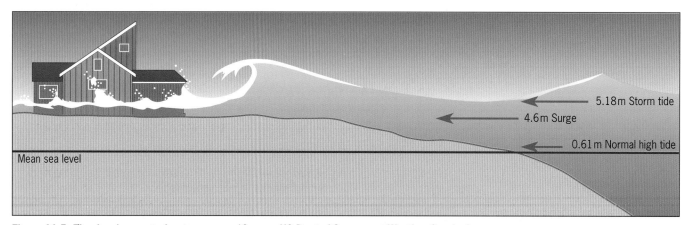

5.18m Storm tide

4.6m Surge

0.61m Normal high tide

Mean sea level

Figure 11.5 The development of a storm surge (*Source:* US Dept of Commerce Weather Service)

?

4   During the 1970s and 1980s, major hurricanes striking the USA were less frequent than in the previous three decades. Despite this, the costs of damage continued to rise. Use Figure 11.6 to explain why this trend occurred.

Damage to property

Hurricanes also lead to widespread destruction of property, especially by wrecking buildings that are poorly constructed (Fig. 11.7). In economically developed areas of the world many buildings do not suffer much more than the loss of windows and roofs. In economically developing countries, though, many houses are less soundly built. Hurricane Gilbert in 1988 left over 500 000 people homeless in Jamaica and the majority had their homes destroyed. The 1970 hurricane in Bangladesh, referred to earlier, left an estimated 85 per cent of the population homeless (Fig. 11.8). Any loose objects, ranging from dustbins to cars, to aircraft, can be tossed around and severely damaged. In addition, power supplies and communications can be disrupted for days.

Figure 11.7 Repairing homes in the Qualey poverty area, Dominican Republic, after Hurricane David, 1979

Figure 11.8 Survivors file across a flooded field seeking relief from the wreckage of the cyclone/tidal wave in which about 5300 died, East Pakistan (Bangladesh), 18 November 1970

?

5a Construct a divided compound bar graph based on Table 11.3.
b In which year were the largest claims made?
c Which disaster made the greatest contribution to these claims?
d Comment on what the graph shows and its implications for insurance companies.
e You could enter the data in Table 11.3 into a spreadsheet on a computer. Put years on one axis and the source of the claims on other. Then enter the value of each claim into the cells in dollars. Enter formulae to calculate total claims for each year.

Effect on agriculture

Crop damage often has devastating consequences in economically developing countries where the economies depend on agriculture. The 1970 hurricane in Bangladesh caused a high tide 7 m above normal during the night. By the morning about 280 000 cattle had been swept away and £40 million of ripening crops had been destroyed. Similarly, Hurricane Gilbert in 1988 destroyed Jamaica's vital export crop of bananas.

Claims on the insurance industry!

Between 1987 and 1993 insurance companies had to pay out more than $50 billion to meet claims (Fig. 11.9). This was following a long list of disasters including Cyclone Iniki, the most powerful storm to hit Hawaii in the twentieth century (Table 11.3).

The claims were so massive that many insurance companies have gone out of business. Since then, the survivors have adopted new approaches: one of these is reinsurance (see Fig. 11.9). The effect of large claims is thus reduced for the smaller company, perhaps in Miami, but can be felt further away in places like London or Tokyo.

Individuals or organisations can insure items of value, like cars or buildings, with an insurance company. An insurance policy is for an amount of money that reflects the value of the item being insured, for example, £50 000 for a building. The client pays a sum of money called a *premium* to the insurance company for this service. Insurance companies make their money by collecting premiums from many clients.

In return an insurance company agrees to pay the client the sum of money insured, £50 000 in our example, when the client claims if the item is stolen, damaged or destroyed. An element of *risk* is involved on the part of the insurance company. The level of premiums has to be set carefully so that the insurance company's income is greater than payments made to clients.

For risks like hurricanes, when claims can be huge, *direct insurers* (the high street insurance companies that consumers are familiar with) spread the risk by *reinsuring* with another larger company or syndicate of individuals, known as '*Names*' to meet claims above a certain amount.

Figure 11.9 The principles of insurance

Table 11.3 Major insurance losses since 1987 (*Source:* Leggett, 1994)

Date and location	Event	Total claim (US$)
October 1987, NW Europe	Unnamed windstorm	2.5 billion
1988, NW Europe	Piper Alpha disaster	1.5 billion*
1989, USA	Exxon Valdez disaster	1.6 billion*
September, 1989	Hurricane Hugo	5.8 billion
1989, USA	San Francisco Earthquake	1.4 billion*
January 1990, NW Europe	Windstorm Daria	4.6 billion
February 1990, NW Europe	Windstorm Herta	1.3 billion
February 1990, NW Europe	Windstorm Vivian	3.2 billion
February 1990, NW Europe	Windstorm Wibke	1.3 billion
July 1990, USA	Colorado storms	1.0 billion
September 1991, Japan	Typhoon Mireille	4.8 billion
October 1991, USA	Oakland wildfire	1.7 billion
August 1992, USA	Hurricane Andrew	16.5 billion
September 1992, USA	Cyclone Iniki	1.6 billion
March 1993, USA	'The storm of the century'	1.7 billion
June-August 1993, USA	Midwest flooding	1.0 billion
October-November 1993, USA	S California wildfires	1.0 billion

*estimate

?

6 Use Figure 11.10 to account for the changing fortunes of Lloyd's Insurance since the late 1980s.

7 Imagine that you own a hotel in Jamaica and that you want to insure it for $10 million.
a Calculate the sum that you would pay to an insurance company in a year when the hotel was hit by a hurricane.
b How do high insurance premiums affect tourism in economically developing countries?

8 Essay: Suggest reasons why economically developing countries often suffer more from the effects of hurricanes than economically developed countries.

9 Outline the main differences between the responses of US and Puerto Rican citizens to hurricanes in Table 11.4. Concentrate on broad patterns and those responses where there are large differences between the percentages.

10 A study by E. Baker and D. Patton in Tallahassee, Florida; Pass Christian, Mississippi; and Galveston, Texas in the 1970s showed that the level of literacy was an important influence on attitudes towards damage prevention measures. If you had responsibility for co-ordinating damage prevention efforts, how would such information affect your approach?

11 Suggest reasons for some of the precautions referred to in Figure 11.11.

```
1986   A typically good year at Lloyd's - profits over $1
billion

1987   Lloyd's profits around $0.7 billion. Lloyd's market
contains more than 400 syndicates and 30 000 Names (wealthy
providers of capital). Total capital in Lloyd's alone is
more than $10 billion.

1988   Lloyd's makes a loss of over $1 billion.

1989   Lloyd's deficit was $3.3 billion.

1990   Lloyd's deficit was $4.3 billion. Profits of British
insurance industry as a whole plunge. Losses from the five
biggest companies total more than $5 billion.

1991   Lloyd's losses expected to top $1.5 billion.

1992   The Lloyd's market contracts to 278 syndicates and
22 000 Names. Insurance cover is withdrawn in the Samoa
Islands in the wake of Cyclone Val, the second supercyclone
to have hit the islands within two years. Catastrophic
losses in the US soar to $23 billion, up from $4.7 billion
in 1991. Nine insurance companies go out of business
immediately following Hurricane Andrew and Cyclone Iniki.

1993   Lloyd's announces an end to unlimited liability for
Names and looks for corporate capital to replace the
market's eroded capacity. The future of Lloyd's hangs in
the balance.
```

Figure 11.10 Record for Lloyd's from the late 1980s

The impact on tourism

Some of the world's major tourist destination areas, like the Caribbean, Mexico and Florida, are prone to hurricanes. However, this does not put off tourists from visiting these areas and warnings should provide ample opportunity for tourists to evacuate. Hurricanes, though, have caused much damage to tourist-dependent economies. In 1993 in the Caribbean, a hotel owner would have to pay an annual insurance premium (see Fig. 11.9) of 2 per cent of the value of the hotel. In addition, if the hotel is hit and damaged by a hurricane, a further 2 per cent of the sum insured (not the value of the claim) has to be paid.

In response to insurance costs, some hotel owners may cancel or reduce their insurance coverage. Consequently, when hurricanes destroy or damage hotels, the owners often find themselves uninsured or underinsured and may become bankrupt. This in turn creates problems for banks and financial institutions that invested money in the hotel. This problem is increasingly affecting areas in the USA like Louisiana, Texas, and Hawaii where property owners have made many claims for hurricane damage.

Perception of hurricanes

There has been little research into people's perception of hurricanes as a hazard, although we might expect perceptions of hurricanes to vary between different areas of the world, particularly between economically developed and developing countries. For a study conducted in the 1970s (Bauman and Simms), a question-naire was given to 360 respondents in three locations in the USA (Tallahassee, Florida; Pass Christian, Mississippi; and Galveston, Texas) and 141 respondents in Puerto Rico (Table 11.4). Although the sample is quite small, as varied a population as possible was interviewed and some important differences in responses emerged.

Table 11.4 Measures of hurricane behaviour (*Source:* Baumann and Simms, 1974)

Sentence stems and completions	Puerto Rico (N=141) (%)	US (N=360) (%)
Before the hurricane		
If a hurricane is predicted, I . . .		
make preparations (unspecified)	55	31
keep on the alert	10	19
feel fear/anxiety	7	4
seek refuge	14	29
[other]	14	17
When a hurricane is coming, I feel . . .		
fear	48	31
anxiety	9	18
concern for the consequences	26	15
desire to take precautions	2	14
[other]	15	22
During the hurricane		
During the hurricane, I . . .		
make preparations (unspecified)	38	17
pray	10	18
communicate with others	13	9
feel fear/anxiety	9	16
protect myself	13	18
protect others	12	8
proceed normally	5	7
[other]	0	7
Going through a hurricane makes me feel . . .		
fear	38	31
anxiety	11	23
negative emotions (unspecified)	6	10
concern for the consequences	36	15
[stays calm]	9	4
[other]	0	17
In a hurricane the people I feel some responsibility for are . . .		
family of procreation (husband or wife and children)	40	37
children	9	17
parents	6	7
family and nonfamily	24	25
nonfamily	14	12
[other]	7	2
After the hurricane		
When a hurricane is over, I . . .		
feel positive emotions	25	26
check results	23	18
thank God	10	10
begin restoration	12	29
aid victims	14	4
feel negative emotions (fear, anxiety)	6	6
[other]	10	7
When a community experiences a hurricane, the feelings among its people . . .		
are of mutual cooperation	31	48
are of fear and anxiety	6	6
are of sadness	42	11
are shared	11	2
are positive	4	17
[other]	6	16

In economically developed countries, posters and literature reassure people and give plenty of advice about precautions to be taken before and during a hurricane (Fig. 11.11). Hurricane warning systems are now so efficient that people have plenty of time to evacuate an area and move inland. The US National Oceanic and Atmospheric Administration estimate that about 80–90 per cent of the population living in many hurricane-prone areas have never experienced the power of a major hurricane. This encourages false impressions of the damaging potential of a hurricane and complacency may then lead to people being reluctant to move to safer areas when a hurricane threatens. This could result in unnecessary loss of life.

When a HURRICANE threatens

KEEP YOUR RADIO OR TV ON... AND LISTEN TO LATEST WEATHER ADVICE TO SAVE YOUR LIFE AND POSSESSIONS

Before the wind and flood

 Fill your car with petrol; check your battery and tyres.

 Have a supply of drinking water. Stock up on foods that need no cooking or refrigeration.

 Have nearby a torch, first-aid kit, fire extinguisher, battery-powered radio.

 Store all loose objects: toys, tools, trash cans, awnings, etc. Board or tape up all windows.

 Get away from low areas that may be swept by storm tides of floods.

During the storm

Stay indoors... Don't be fooled if the calm 'eye' passes directly over... And don't be caught in the open when the hurricane winds resume from the opposite direction.	Listen to your radio or TV for information from the Weather Bureau, civil defence, Red Cross, and other authorities.

After the storm has passed

Do not drive unless necessary. Watch out for undermined pavement and broken power lines.	Use extreme caution to prevent outbreaks of fire, or injuries from falling objects.	Report fallen power lines, broken water or sewer pipes to proper authorities or police.	Use phone for emergencies only. Jammed switchboards prevent emergency calls going through.

Figure 11.11 Advice in the event of a hurricane (*Source:* US Dept of Commerce, Weather Bureau, 1963)

Hurricane Andrew, Florida, USA

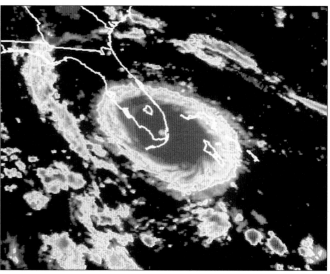

In mid-August 1992 a hurricane was building up in the middle of the Atlantic that was to become one of the most powerful in the twentieth century. On the afternoon of 23 August it reached the Bahamas, causing widespread damage with unconfirmed reports of four deaths in the outlying islands. Satellites tracked the hurricane as it moved north-east towards the USA off the east coast of Florida (Figs 11.12–11.13). The US National Hurricane Centre and the media issued warnings that this would be a storm on a scale not seen in Florida for 50 years. Wind speeds of over 240 km/h, along with torrential rain and a tidal surge of 5m were predicted.

In response to the warnings, the Florida population frantically stockpiled tinned food and batteries, and sales of plywood soared as people battened down their windows. Any property, like cars or yachts in Palm Beach, that was likely to move in the strong winds was secured or stored away. Approximately one million people evacuated their homes near the coast and moved inland to higher ground, designated shelters or the homes of relatives.

Figure 11.12 NOAA satellite image of Hurricane Andrew, 24 August 1992. The colours approximate to cloud thickness on a scale from red (thickest), pink, yellow (the eye), blue to black (clear sky).

Figure 11.13 Weather chart for Hurricane Andrew off the east coast of Florida (*Source*: NOAA)

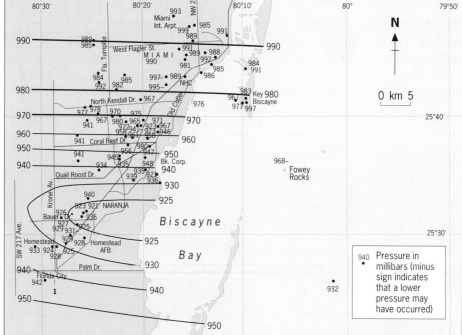

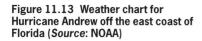

12 What does the tiny eye to Hurricane Andrew visible in Figure 11.12 suggest about the hurricane? Explain your answer.

13a State how the weather chart (Fig. 11.13) indicates the position of the cyc.
b How does the weather chart show the intensity of Hurricane Andrew?

The impact of Hurricane Andrew

Early on 24 August, Hurricane Andrew slammed into south Florida. Gusts of up to 270 km/h were recorded as the hurricane pounded the area for about eight hours. The damage caused was extensive (Figs 11.14–11.15) making it the most devastating hurricane in Florida since 1937. At least 12 people died and hundreds were taken to hospital with injuries. More than 50 000 homes in southern Florida were destroyed and two million people were left without electricity. There were also several cases of looting. Although the National Guard successfully caught many looters, the police were too busy to stop all scavenging.

Hurricane Andrew tracked across the Gulf of Mexico gathering strength over the warm sea. About 300 000 people fled New Orleans and a further 500 000 from the

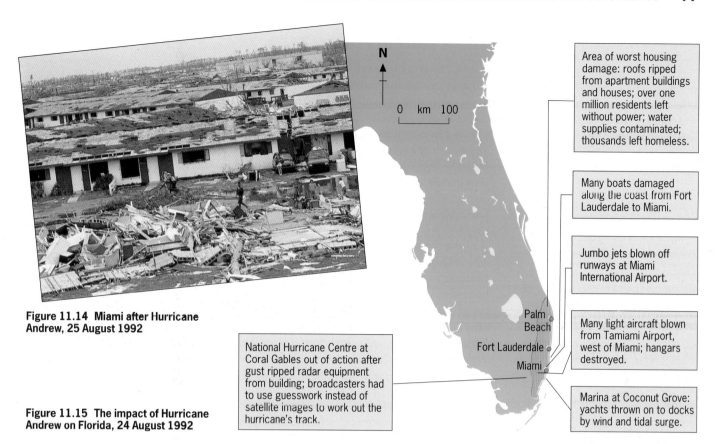

Figure 11.14 Miami after Hurricane
Andrew, 25 August 1992

Area of worst housing damage: roofs ripped from apartment buildings and houses; over one million residents left without power; water supplies contaminated; thousands left homeless.

Many boats damaged along the coast from Fort Lauderdale to Miami.

Jumbo jets blown off runways at Miami International Airport.

Many light aircraft blown from Tamiami Airport, west of Miami; hangars destroyed.

National Hurricane Centre at Coral Gables out of action after gust ripped radar equipment from building; broadcasters had to use guesswork instead of satellite images to work out the hurricane's track.

Marina at Coconut Grove: yachts thrown on to docks by wind and tidal surge.

Figure 11.15 The impact of Hurricane
Andrew on Florida, 24 August 1992

low lying areas near the coast in other parts of the state of Louisiana. Again, extensive damage was caused as winds of up to 225 km/h affected the area on 26 August. A further three deaths were reported in Louisiana. The hurricane passed on towards Texas and Alabama before slowly dying out.

People's initial response was one of shock and disorientation. Thousands of people were living in their cars or makeshift dwellings. Families were forced to scavenge for food and clean water. Although President Bush declared south Florida a national disaster area, relief efforts were delayed by red tape and confusion. Within hours, tent cities and 20 mobile kitchens for over 4000 people should have been set up. Only by the weekend after the hurricane had four kitchens been established, though, and people were outraged at this slow response to the disaster. Eventually, about 8000 soldiers were landed while many military flights delivered portable toilets, tents, generators, water purifiers and earth-moving equipment – making this relief effort one of the biggest in American peacetime history.

Insurance claims

Hurricane Andrew was also one of the costliest disasters in American history. Allstate Insurance, a subsidiary company of the Chicago based Sears-Roebuck, was expecting over 150 000 claims relating to hurricane damage. This would have the effect of giving Sears-Roebuck a loss amounting to several hundred million

dollars for the quarter. Allstate had to bring in an additional 1100 employees from out of the state of Florida to process the claims. Eight insurance companies in Florida were pushed out of business while many others were forced to put a limit on future policies. Total insurance claims were eventually put at $18 billion with over $7 billion coming from the state of Florida.

Following Hurricane Andrew, many companies have cancelled or limited property insurance in coastal areas of Florida. Thousands of property owners and businesses have been left without cover. For those left with insurance, in many cases their premiums have risen by as much as 40 per cent.

People also learnt much from the hurricane. The powerful winds showed up poor building and inadequate policing of construction work. Concrete block construction fared much better than buildings with timber frames and this will influence people's future house purchasing behaviour. Tighter building regulations have since been introduced. Hopefully, the inefficiencies that led to such a slow initial response to the disaster by the government and the military will not be repeated in future.

14 Attempt to classify the different impacts of Hurricane Andrew. Consider both positive and negative aspects of the hurricane.

Insurers could certainly do much to facilitate the introduction of new non-polluting technology thereby helping to achieve a stable global situation in which they would be better able to predict the risks they would be facing – risks that were contained at a level that made underwriting a science rather than a lottery.

Figure 11.16 Quote from Lloyd's in-house magazine (*Source:* Peter Corbyn, *One Lime Street*)

15 One consequence of global warming is a rise in sea level. Explain how this would increase the damage caused by hurricanes. Use diagrams to clarify your answer.

16 Essay: Study Figure 11.16. The author refers to a number of steps that insurance companies could take to 'achieve a stable global situation in which they would be better able to predict the risks they would be facing'. How could the development of alternative energy sources influence hurricanes in the future and benefit the insurance industry?

11.5 A stormier future?

Global warming and more intense hurricanes

Many scientists have suggested that global warming (see section 4.4) will be likely to increase both the severity of hurricanes and the number of places where they occur. Warmer seas could mean that many more areas of the Pacific could be affected by hurricanes. This may mean that people will leave some tropical coastal areas as they become more hazardous. In Australia, meteorologists estimate that an increase in average sea surface temperatures of about 2–3°C would result in hurricanes affecting areas as far south as Brisbane and with greater strength. It is also possible that more of Japan could be affected, the north-east coasts of the USA, parts of New Zealand and even southern Europe.

An uncertain future for insurance companies

The insurance business, as we have seen, is all about assessing risk (see Fig. 11.9). Concern over global warming, making the problems of insurance companies worse, has led to many efforts to predict claims and limit the crippling losses that can occur as a result of major storms. Some of the largest companies have started to employ or consult their own meteorologists, while others have set up special 'greenhouse teams'. One example is after Hurricane Hugo, in 1989, when a Lloyd's syndicate approached a leading British meteorologist who said that the southern USA was likely to experience more storms in the future. The syndicate took note of this advice and started to reduce its commitments in the region – including Florida. Many companies noted this foresight with envy after Hurricane Andrew occurred in 1992. Perhaps there is a future employment opportunity here for you!

The insurance business is one of many that are becoming aware of the need to work with environmental groups and governments to consider issues affecting global warming (Fig. 11.16). Although rather alarmist, one nightmarish prediction involves super-hurricanes with wind speeds of up to 200 km/h. Consider the effects of such a hurricane in a heavily populated area like New York or Tokyo.

Summary

- Hurricanes have caused more deaths than any other form of natural disaster, except perhaps droughts. The effects of a hurricane can be catastrophic, causing widespread damage to property, disrupting communications and causing loss of life.
- Hurricanes have several different names in various parts of the world.
- Hurricanes are intense storms with wind speeds that average over 119 km/h.
- Hurricanes are largely restricted to the tropics where the right combination of conditions exists for their formation.
- The insurance industry has been seriously affected in recent years by massive claims following several large hurricanes.
- Hurricanes are particularly serious for economically developing countries where the cyclones frequently wipe out valuable export crops and also damage the tourist industry.
- Many scientists fear that, if global warming continues, there will be an increase in the frequency, severity and spatial distribution of hurricanes.

12 Local climates

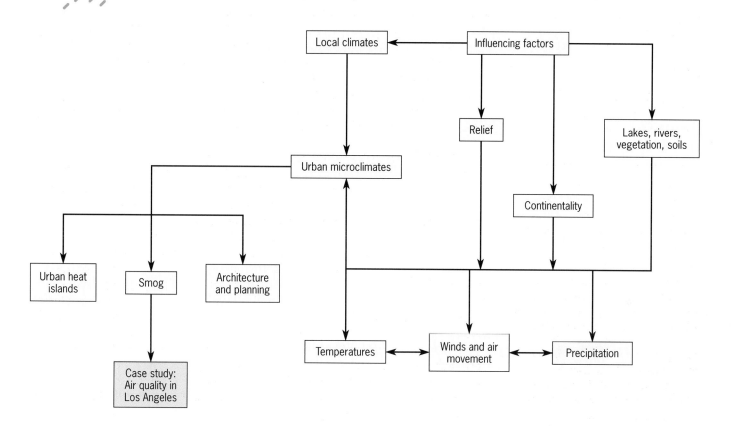

12.1 Introduction

Local **climates**, or **microclimates**, relate to a small area and possess clear contrasts with other nearby areas. Such contrasts may result from the effects of buildings, differences in the direction in which slopes face (**aspect**), proximity to water or the nature and density of vegetation cover. The space which meteorologists study can be large or small. At one extreme, a boulder can generate its own microclimate. Plants may grow on the sheltered side and be absent from the exposed side. This could result from differences in wind speeds and amounts of light and **insolation**. At the other extreme, urban microclimates have received much attention in recent years. They involve the study of the effect the built environment has on atmospheric processes and includes air pollution. The scale of such studies could involve many tens of square kilometres.

Localised changes in the environment can lead to pronounced changes in atmospheric temperatures, wind speeds, **humidity** and **precipitation**. These, in turn, can affect the quality of the air by allowing a build-up of pollutants that do not readily disperse (Fig. 12.1). Many of you will know how unpleasant it can be to breathe in traffic fumes in a busy street. This will generally make us turn our heads away or perhaps cough. However, for people with breathing difficulties or hay fever sufferers, severe discomfort can result in extreme cases requiring medical attention.

In addition, the study of microclimates is of importance to many other groups of people including planners, architects and farmers.

Figure 12.1 Tokyo oxygen bar, Japan

?

1a People have expressed concern recently about the alarming increase in asthma cases. What does Figure 12.2 suggest is the link between these cases and the weather?

b If a link between asthma and traffic is proven, suggest how this should influence government transport policy.

Number of asthmatic children doubles

A child is admitted to hospital every 10 minutes in Britain because of asthma, a campaigning charity said yesterday.

The National Asthma Campaign said the number of cases had doubled over the last 15 years and one in seven school children now suffered from the illness.

Melinda Betts, the charity's executive, said although there were many theories about the rapid spread of asthma 'the truth is that no one knows why there has been such a hugh explosion'.

The timing of exposure to early child-hood infections, the dung of house mites building up in the air inside modern, well insulated homes and decreasing quantities of fresh fruit and vegetables in diets are all thought to be involved in the onset of the illness.

Air pollution, both from traffic fumes outside the home and possibly gas stoves within, can also worsen symptoms.

The National Asthma Campaign called for statutory air quality standards to be drawn up by the government and rapidly enforced across the country.

Figure 12.2 Air quality and asthma (*Source: The Independent*, 11 October 1994)

12.2 Urban microclimates

One of the most distinctive and varied of all local climates is that of towns and cities (Fig. 12.3). The changes to the natural environment provide an example of *positive feedback*. Human activities disturb the system to the extent that some processes are exaggerated, such as the increase in temperatures. Other processes are reduced, for example, higher temperatures tend to encourage lower relative humidities in urban areas. The built environment changes many aspects of the natural environment, particularly roughness, thermal properties and hydrology. These combine to create an urban microclimate that can have important social, economic and ecological effects.

Temperatures

One of the clearest differences between towns and cities and the surrounding countryside is that they are generally warmer – creating a **heat island.** In most mid-latitude European or North American cities, average minimum winter temperatures are 1–2°C higher than in the rural surrounds. This results from a number of characteristics of the urban environment.

• Lower wind speeds in comparison with rural areas allow warmth to accumulate. The fact that the heat island effect can still be measured in higher wind speeds reflects the importance of human modifications to the natural environment.

• Buildings have a higher capacity to retain and conduct heat than soil and vegetation: windows let in sunlight that is absorbed by dry surfaces. Heat is released more slowly from the mass of buildings and artificial surfaces. Urban temperatures consequently lag behind the **diurnal** and seasonal changes.

• The burning of fossil fuels in homes, offices, industry and by transport is one of the major sources of heat. Energy from heating systems in some large north European cities has reached 25 W/m² during the winter – this exceeds the local area *inputs* of energy from the sun.

• **Smog** and pollution traps outgoing radiant energy and this can help to maintain higher urban temperatures.

There can also be localised differences in temperatures within the urban environment resulting from exposed and shaded areas. This is most effective where there are many tall buildings and at times of the year when the angle of the sun's rays is low.

Figure 12.3 Urban landscape, San Francisco, USA

2 Although the heat island effect can be measured during the day, it is often strongest at night. Try to account for this.

3a Study Figure 12.4. Describe the pattern of temperatures over Christchurch.
b Use Figure 12.5 to account for the peaks of temperature at locations A and B.
c Use Figure 12.5 to account for the lower temperatures at the locations lettered C and D.

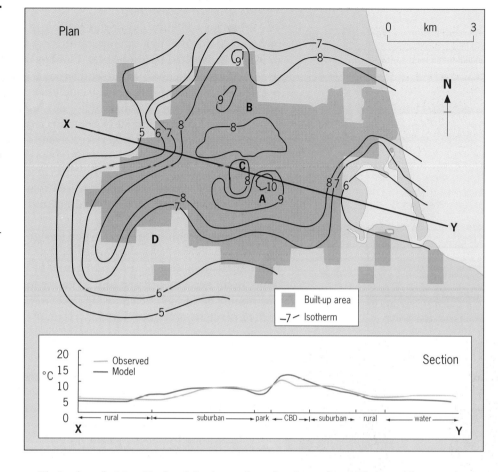

Figure 12.4 The urban heat island of Christchurch, New Zealand

Christchurch, New Zealand, is situated on the Canterbury Plains and covers an area of about 140 km². It is ideal for a study of the urban heat island effect (Fig. 12.4) because of the flatness of the area (relief does not interfere with the distribution of temperatures) and simple pattern of land uses (Fig. 12.5).

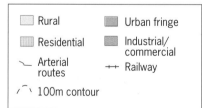

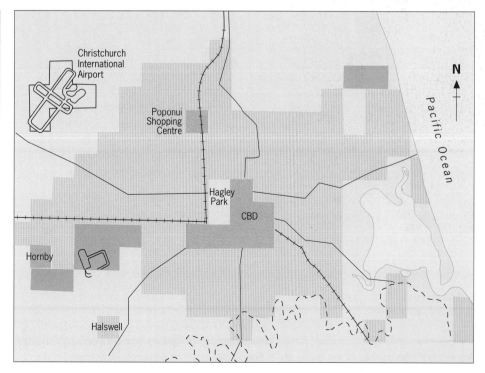

Figure 12.5 Land use map of Christchurch, New Zealand

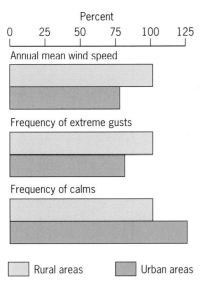

Percent

Annual mean wind speed

Frequency of extreme gusts

Frequency of calms

Rural areas   Urban areas

Figure 12.6 A comparison of selected wind speed characteristics between urban and rural areas

Winds and air movement

The urban heat island effect seems capable of producing its own winds. Studies have shown that the higher temperatures can lead to lower **pressure** over cities that draws air in from the surroundings. This has been noted in a number of north American cities such as Toronto and St Louis (Auer, 1978). These winds can be responsible for preventing a heat island effect in smaller towns when windspeeds are greater than about 20 km/h.

A city has its own relief (Fig. 12.3), and this influences wind speeds which therefore differ considerably between urban and rural areas (Figure 12.6). Wind speeds are also spatially more variable since the winds are concentrated between buildings. This can lead to powerful gusts in some streets that are roughly parallel to the wind direction. Consequently, streets at approximately right angles to the wind direction are sheltered by buildings and remain comparatively calm.

Wind speeds, architecture and urban design

Many of us will have experienced walking comfortably along a street between tall buildings to then turn a corner and nearly be blown off our feet by a strong gust of wind. Architects and urban planners clearly need to consider wind speeds when positioning buildings and choosing street or path layouts. For example, the campus at the University of Surrey (Fig. 12.7) is on the side of Stag Hill near the centre of

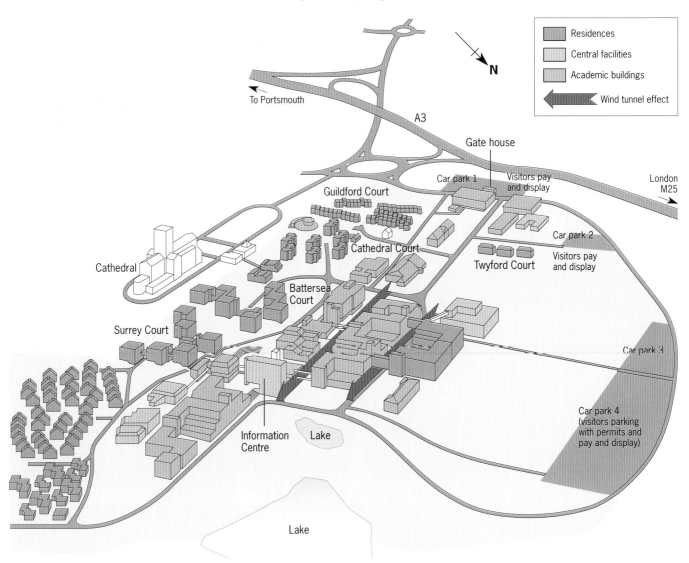

Figure 12.7 The campus of the University of Surrey (*Source:* The University of Surrey)

4 Describe and account for the wind speed characteristics shown in Figure 12.6. Use Figure 12.3 to help you.

5 Study Figure 12.7. Suggest a simple solution to the problem of concentrated winds along the main routes east-west across the campus where the wind tunnel effect occurs.

6 Compile a short list of 'do's' and 'dont's' for planners to reduce wind-tunnel effects based on Figure 12.7.

Guildford. The plan reflects a number of good features, such as pleasant views from the halls of residence, and the plan's simplicity makes it easy to memorise. However, on windy days strong gusts follow the contours of Stag Hill and so are concentrated along the 'valleys' between the tall buildings. As a result, it is often difficult to walk along these parts of the campus. In contrast, the pathways along the sides of buildings are sheltered and wind speeds are very low.

The pressure of winds on buildings is another important consideration, for buildings have to be able to withstand the maximum likely wind speeds. These have to be measured not at ground level, where there is more friction, but at the level of upper floors.

Precipitation

The heat island effect that we have already studied also influences precipitation. This is because the higher temperatures encourage lower pressure over cities which draws air in from the surroundings. This then leads to upward air movement that can in turn encourage **convectional** rainfall. This can be further enhanced by the **orographic**, or relief, effect of tall buildings. In addition, cities also generate huge quantities of dust and particles which, once in the atmosphere, form a dome over the urban area. Particulate matter can be 3–7 times greater over cities than surrounding rural areas, and hold up to 200 times more gaseous pollutants, e.g. sulphur dioxide and carbon monoxide. Particulate matter can act as **hygroscopic particles** (see section 7.5) that provide nuclei for **condensation.**

The METROMEX (Metropolitan Meteorological Experiment) studies of St Louis in the late 1970s represent one of the most detailed analysis of urban microclimate to date. They showed that there was a higher incidence of cumulus cloud development over the city, particularly late in the day, while summer rainfall totals were 20 per cent higher than in the surrounding rural areas. This was accounted for by a high concentration of aerosol particles (**condensation nuclei**) and instability associated with higher urban temperatures. The upward-moving air currents carry the aerosol particles to cloud base level.

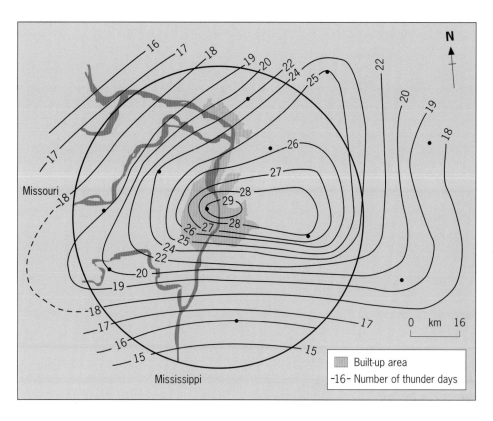

Figure 12.8 Average of summer thunder days (1971–5), St Louis, USA

7 Describe and account for the pattern in Figure 12.8.

8 Account for the differences for the rural and urban areas shown in Table 12.1

9 How does Figure 12.9 suggest that the higher incidence of thunderstorms in St Louis affects human activities?

10a Use Figure 12.10 to describe and explain how amounts of insolation received by north and south facing slopes differ in the northern hemisphere.
b How is the receipt of insolation on north and south facing slopes affected by latitude and the gradient of slopes?

Table 12.1 Urban related increases in thunderstorm activity at St Louis, USA

| | Frequencies | | | Percent increase $\left(\frac{U-R}{R} \times 100\right)$ |
	Urban	Rural	Urban–rural	
Thunder days	26.8	18.5	8.3	+45
Thunder periods	33.6	23.1	10.5	+45
Duration of thunder (all values in minutes)				
Thunder periods	145	93	52	+56
Very light thunder rates (≤5)	49	44	5	+11
Light thunder rates (6–11)	64	48	16	+31
Moderate thunder rates (12–60)	56	46	10	+22
Intense thunder rates (>60)	86	47	39	+83

Thunder days = summer average values
Thunder periods = discrete periods separated by >1 hour from other periods
Thunder rates based on number of peals per hour (values in brackets)

Not only were precipitation totals higher but hail stone size and energy and thunderstorm activity were also greater than for rural areas (Table 12.1). The METROMEX studies generally showed that the urban environment affected all synoptic **weather** conditions. The urban environment makes the precipitation process more efficient and active in any kind of unstable conditions. However, the findings suggested that urban mechanisms tend to intensify rather than start convective activity.

It must be stressed that St Louis is in a continental interior location that typically experiences strong convectional activity, particularly in the summer months. This obviously influences the features of its microclimate. Nevertheless, the features of the urban microclimates that we have studied are found in cities all over the world. However, we must also take into consideration local conditions such as relief and proximity to the sea, as well as the climate of that part of the world, as these will be reflected in the microclimatic conditions.

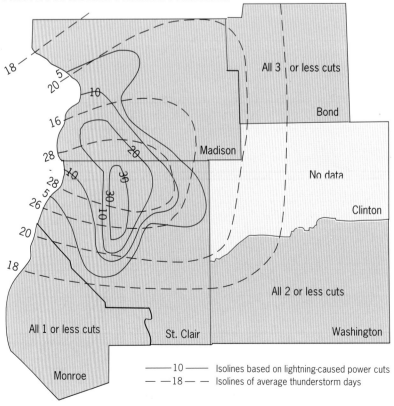

Figure 12.9 Pattern of summer lightning-caused power failures with average thunder days

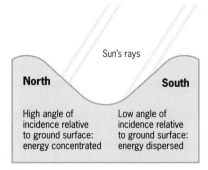

Figure 12.10 The influence of aspect in the northern hemisphere on the angle of incidence of the sun's rays

11 Study Figure 12.11.
a Describe and explain how aspect has influenced the distribution of land uses in the valley around Scuol in south-east Switzerland.
b If you wanted to plant vineyards, where in the valley would you site them? Explain your answer.

12 Suggest how knowledge of microclimates is useful to farmers living in mountainous areas.

12.3 The influence of relief

Temperatures

We have seen previously that there is a decrease in temperature with altitude of about 6.5°C for every 1000 m in the **troposphere** (see section 7.6). It follows, then, that air temperatures will generally be lower in upland and mountain areas compared with surrounding lowlands. This is because the air is thinner at high altitudes and air pressure lower, so temperatures fall quite rapidly above the ground surface. As air is heated from the ground and there is often very varied relief in upland areas, temperatures can also vary over quite small distances.

One factor that can influence the radiation budget in upland areas is aspect – the direction in which a slope faces. This influences the angle of incidence of the sun's energy (Fig. 12.10). This uneven exposure to the sun means that some slopes receive much more light as well as warmth. This has an important effect on the distribution of plant species in upland areas and also on human activities (Fig. 12.11).

Temperature inversions and air quality

A number of circumstances occur when the normal fall in temperature with altitude in the troposphere is reversed, known as a **temperature inversion**. Such an irregularity involves warmer air lying above colder air. As colder air is denser, it cannot escape and therefore very stable atmospheric conditions result. Temperature inversions are consequently associated with clear skies, little air movement and, depending on the latitude, high temperatures.

Temperature inversions commonly occur in upland areas at night (Fig. 12.12). They can lead to the development of **radiation fog** (see section 8.3). As the sun rises during the morning, although the bright upper surface of the fog initially reflects some insolation, rising air temperatures mean that the fog will gradually evaporate and disperse.

Temperature inversions can also take place over flatter areas and are common over extensive lowlands and basins in all the large continental areas of the northern hemisphere. They occur particularly in the winter when there are persistent **anticyclonic** conditions. Clear skies at night result in the loss of radiant heat from

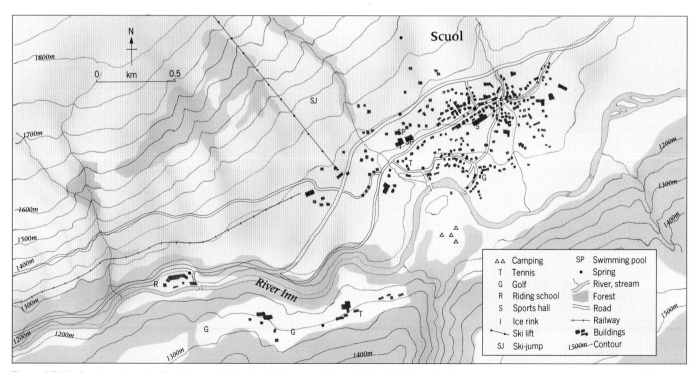

Figure 12.11 Land use for the Scuol area, Switzerland (*Source:* National Atlas of Switzerland)

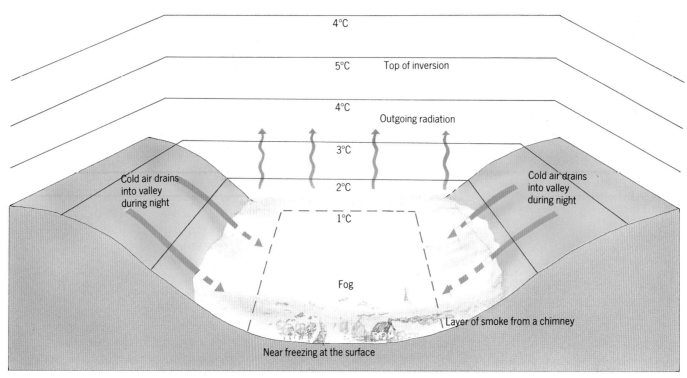

Figure 12.12 Features of a temperature inversion

Figure 12.13 Coffee bushes planted part-way down the side of a valley to protect them from frost damage, southern Brazil

the surface, and the low angle of incidence of the sun's rays also results in low temperatures. In addition, persistent cloud cover can also block out much of the sun's energy. The ground becomes very cold and drops to temperatures below the air temperature. This results in the air adjacent to the ground being colder than air above. The normal **lapse rate** may not be regained until an altitude of several hundred metres.

Since temperature inversions are associated with stable atmospheric conditions, they can lead to hazardous air quality conditions. This is because pollutants are not easily dispersed horizontally while there is little air movement. There is also an absence of vertical dispersal because cold air underlies the warm air.

Figure 12.14 shows a how a temperature inversion can lead to severe ground level pollution. It is particularly important to know when inversions are likely to occur in industrial areas because they trap pollution near the ground and, if the source continues, the concentration can increase. High concentrations of particulates in the air can encourage the formation of smog (Fig. 12.15). This is a combination of fog and smoke or pollution that occurs when condensation takes place on the particles that act as hygroscopic nuclei.

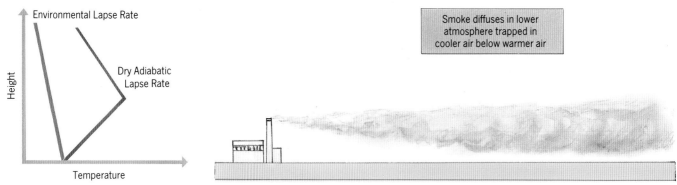

Figure 12.14 Pollution trapped near the ground by a temperature inversion

13 Study Figure 12.13. Suggest why coffee bushes have only been planted to halfway down the side of the valley.

14 With the aid of a diagram similar to Figure 12.14, explain how a change in chimney height could reduce the problem of pollution near the ground.

Smog is a very severe meteorological hazard because it concentrates pollutants in the air that we breathe. Such conditions have caused numerous deaths in industrial regions, such as in Pennsylvania in 1948, which was then an area of steel works and heavy engineering. Similarly, over 12 000 people, particularly the young and elderly, died from respiratory failure during the very severe smogs in London during 5–9 December 1952. The cause was a temperature inversion over the city when visibility fell to less than 10 m over a period of 48 consecutive hours. This led to the Clean Air Act (1956) that put strict controls on pollution in UK cities. These included restrictions on industrial emissions and the introduction of smokeless zones where the public are not allowed bonfires and, more importantly, have to use smokeless fuels in their homes.

Air quality in Los Angeles, USA

The physical background
The city of Los Angeles is in a huge basin covering over 4000 km² (Fig. 12.16). This part of California frequently experiences temperature inversions that lead to severe smogs (Fig. 12.15). The inversions result from westerly air streams which cross the cold California current. As they enter the basin, they are trapped below warmer air. This particularly happens during the day when there is strong subsiding air, whereas nights are cooler and skies early in the morning are comparatively bright and clear. Also, Los Angeles does not experience strong winds very frequently which would otherwise help to disperse smog.

Air quality in the Los Angeles basin
The main source of pollution is transport. Los Angeles is renowned for its freeways and very high levels of car ownership where over 5 million cars burn about 40 million litres of petrol each day. Car exhausts are a cocktail of nitrogen oxide (NO), carbon monoxide (CO), unburnt petrol vapours and hydrocarbons. Sunlight then converts nitrogen oxide and hydrocarbons into nitrogen dioxide (NO_2) – a chemical reaction that also releases fine particles into the air. These particles can also encourage condensation and lead to photochemical smog.

Photochemical smog not only smells unpleasant but can also cause eye irritation, coughing and tiredness. While smog causes great discomfort, especially for people who suffer from asthma or bronchial complaints, it can also lead to death. Consequently, authorities carefully monitor pollution levels in Log Angeles and, in severe conditions, they have the power to close factories and even bring traffic to a standstill on the freeways.

Figure 12.15 Smog sits at ground-level over Los Angeles, USA

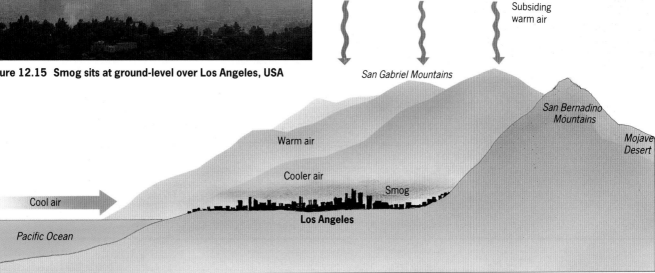

Figure 12.16 Smog in the Los Angeles basin

Los Angeles

?

15 Use Figure 12.16 to explain why pollution has difficulty escaping from the Los Angeles area.

16 Explain how anticyclonic conditions could encourage smog in Los Angeles.

17 Study Figure 12.17. You are a member of the LA public health authority. Devise a plan to reduce particles and gases which encourage smog.

Figure 12.17 Small particles in cigarette smoke contribute to smog (*Source: New Scientist*, 20 Aug. 1994)

LIKE millions of tiny chimneys, the cigarettes puffed by Los Angeles smokers every day make a small but detectable contribution to the thick smog that often shrouds the city. According to Glen Cass, an evironmental engineer at the California Institute of Technology, cigarette smoke makes up 1 to 1.3 per cent of the fine particles polluting the air around Los Angeles.

Although fine particles are present in smaller quantities than gaseous pollutants such as carbon monoxide and ozone, they are responsible for almost all of the visible haze that robs Angelenos of what were once breathtaking views. 'When you can't see the mountains any more, you have an overloading of small particles in the atmosphere,' says Cass.

The study by Cass and his colleagues is part of a major research programme to track the sources of particles polluting LA's air. Although cigarette smoke makes only a small contribution to the total, it is one of a half-dozen or so minor sources that collectively account for about a quarter of the pollution. Even if it proves impractical to control these minor sources, says Cass, planners need to take note of them when calculating the effect of regulating emissions from more important sources such as vehicle exhausts, fireplaces and barbecues.

Winds and air movement

At very high altitudes winds are often very powerful (only a few of the highest mountain peaks interrupt the flow of air). This occurs because air moving across extensive upland areas is forced to concentrate vertically which also leads to acceleration, for example, at the top of Mt Everest winds of up to 320 km/h have been recorded. Conversely on leeward slopes, the air diverges and slows as it spreads out. At lower altitudes every hill, valley and variation in relief represents friction and a disturbance to the natural air flow (Fig. 12.18).

Winds will clearly accelerate as they move through narrowing valleys or mountain passes and the mass of air is forced to constrict. One example of this is when high pressure lies to the north of the Alps and low pressure over the Mediterranean. Air is then drawn from the mountains southwards which often leads to the Mistral. This is a strong, gusting wind that becomes concentrated in the Rhône Valley on its way to the south of France.

Similarly, winds will accelerate around a hill and then slow as they spread beyond the hill and as friction takes effect. This can influence air quality in mountain areas as a result of pressure differences either side of a hill (Fig. 12.19). It follows that in areas of very varied relief there can be considerable local variations in wind speeds. Valleys parallel to prevailing air streams may experience high velocities while valleys at right angles will be sheltered and have low wind speeds.

Figure 12.18 Oblique aerial view of an alpine scene

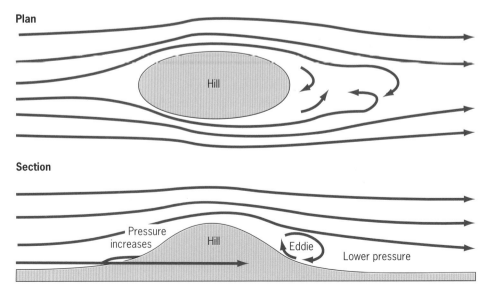

Plan

Section

Figure 12.19 Changing velocity and behaviour of air as it passes over and around a hill

Figure 12.20 Formation of anabatic winds

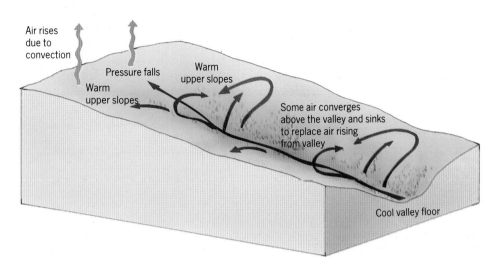

Anabatic and katabatic winds

Mountain areas also experience diurnal wind reversals similar to land and sea breezes (Fig. 12.20). They are usually shallow winds moving at slow speeds of about 15 km/h.

Fohn winds

Flows of air over mountains can also create a **fohn** effect (Fig. 12.21). When this effect occurs it can result in rapid increases in temperatures and falls in **relative humidity** with a number of consequences. When the fohn wind blows down the northern slopes of the Alps in the spring, it can lead to rapid snow-melt and flooding and can also trigger avalanches. The Chinook in North America is another fohn wind that occurs on the eastern slopes of the Rockies. The drying effect of the wind here can encourage the outbreak of fires in the dense coniferous forests.

Precipitation

Mountain areas tend to be wetter than lowlands. They are often shrouded in cloud or mist and are among the wettest places in the world. Mount Wai-'ale-'ale in Hawaii is shrouded by moist clouds for 354 days a year and experiences over 11 600mm of rain each year. The wettest place in England is Seathwaite Farm, Borrowdale, in the heart of the mountains of the Lake District. Much of this rainfall is orographic, or **frontal**, rain enhanced by the influence of relief (see section 8. 3).

Cloud development and height are influenced both by relief and the time of year.

Figure 12.21 The cause of fohn winds

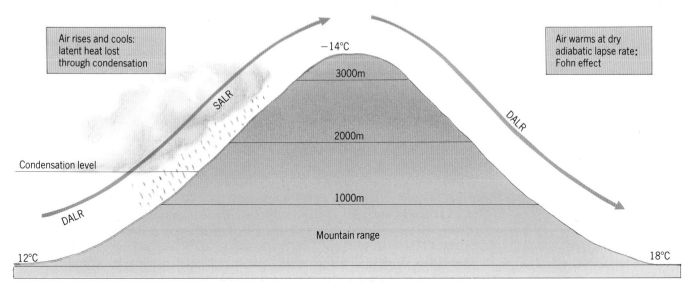

18a Use Table 12.2 to plot a graph for precipitation above a copy of the transect across the Alps in Figure 12.23.
b Describe and offer a general explanation for the curve on your completed graph.
c How is this knowledge useful to mountaineers?

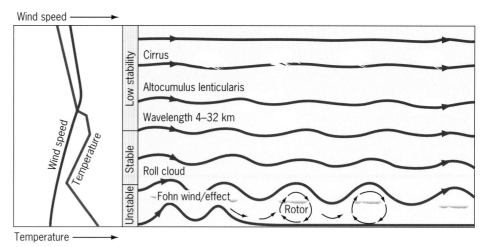

Figure 12.22 Waves in air currents and associated weather downstream from a mountain range

Table 12.2 Average annual precipitation for selected districts in the eastern Alps

District	Mean elevation (metres)	Precipitation (annual total in mm)
North foreland	670	970
North edge	768	1550
Central Inn	809	899
Otztal Alps	1619	767
Bolzano	809	700
South-east Dolomites	1315	1211
Trentino	236	950

Although upper slopes frequently experience heavy rain, there is an upper limit beyond which a lack of water vapour restricts cloud development.

Thus, rainfall and cloud development are higher in the day than at night and in the summer in comparison with the winter.

Mountain ranges also establish waves in winds that in turn affect cloud development and the distribution of precipitation (Fig. 12.22). Precipitation can therefore vary considerably in mountain areas over short distances. Valleys frequently receive less rainfall than upper slopes although much drains to the valley floors.

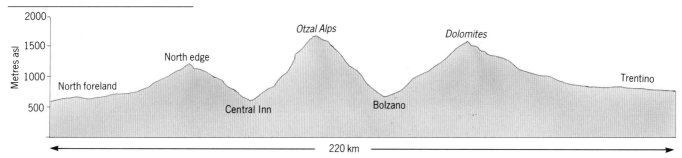

Figure 12.23 Transect across the Alps

12.4 The influence of continentality

Another major influence on the weather and climate of a place is its distance from the sea, or the influence of **continentality**. Figure 12.24 shows climate graphs for two Canadian cities representative of a maritime and continental interior location. There are a number of clear differences that can be accounted for by the influence of continentality.

Temperatures

The sea has a high **thermal inertia**. This means that it is slow to respond to the seasonal changes in insolation. The sea, in other words, warms and cools slowly. In contrast, land has a low thermal inertia so it warms and cools quite rapidly.

Winds and air movement

Coastal resorts are often quite windy places. This can be invigorating for holidaymakers and helps to disperse pollution and maintain good air quality. The sea provides a smooth surface and therefore little friction to slow down winds. In addition, temperature differences between land and sea areas (thermal inertia) generate local land and sea breezes (Fig. 12.25). These are irregular in mid-latitudes but occur virtually every day in the warmer tropics.

19a Study Figure 12.24. Describe the differences between the climate of Vancouver and Winnepeg
b Explain which differences can be accounted for by distance from the sea.

?

20a Use arithmetic graph paper to draw a scattergraph based on the data in Table 12.3. Label each point on the graph by its location reference number. Label the axis and plot a best fit line. Give the graph a title.

b Use Spearman's rank correlation to test the statistical relationship between the same two sets of data. Test the validity of your result by using significance tables.

c Comment on the significance of your result and explain the relationship it suggests.

d Comment on the strengths and weaknesses of the two methods you have used to analyse the data.

21 Study Table 12.4. Why do the locations have such different winter temperatures although they are at virtually the same latitude?

22 Draw a similar diagram to Fig. 12.25 showing the reverse circulation that occurs at night.

23 Suggest why land and sea breezes are more frequent during anticyclonic conditions.

24 How would an understanding of land and sea breezes be of value to glider pilots near the coast?

25 Research some coastal resort data, e.g. changes in wind direction. Use polar graph paper to show directions of winds over time.

Table 12.3 Annual temperature ranges and distance from the sea for selected European cities

Location	Annual temperature range (°C)	Distance from the sea (km)
Birmingham	12	107
Brussels	16	100
Bucharest	27	200
Cherbourg	11	2
Hamburg	18	80
Lisbon	11	8
Lyon	19	275
Madrid	18	325
Monte Carlo	14	1
Munich	21	300
Paris	15	180
Santander	9	1
Venice	19	1
Vienna	23	350
Vlissingen	15	2
Warsaw	23	275

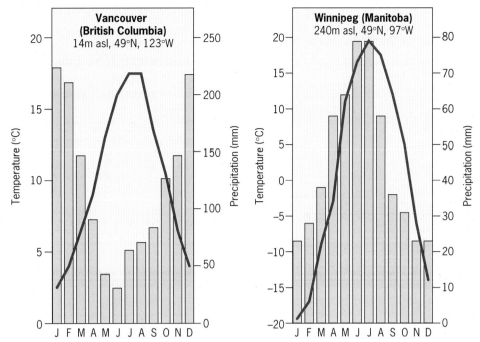

Figured 12.24 Rainfall and temperature for two Canadian cities

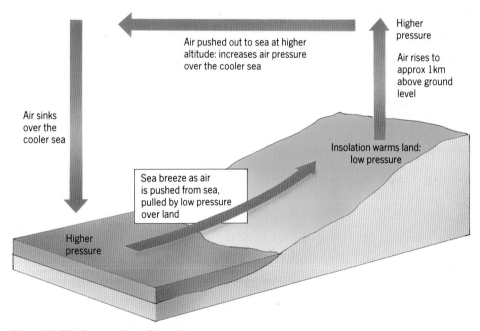

Figure 12.25 Causes of sea breezes

Table 12.4 Annual monthly temperature statistics for Bordeaux, France and Portland (Maine), USA

	Jan	Feb	Mar	Apr	May	Jun	Jul	Aug	Sept	Oct	Nov	Dec
Bordeaux 46 m asl, 44°50'N												
Temp (°C)	5.5	6.5	9.5	11.5	14.5	18	19.5	20	17.5	13.	9	6
Portland 31m asl, 43°40'N												
Temp (°C)	–5	–4.5	0.5	6	12	17	20	19	15.5	10	3.5	3.5

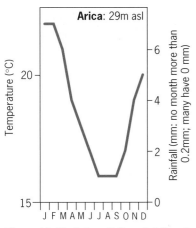

Figure 12.26 Arica, Chile: rainfall and temperature

?

---

26 Look at Figures 12.26, 6.8–6.9 and a map of Chile, South America, in an atlas. Draw an annotated cross sectional diagram on an east-west axis that accounts for the aridity of Arica and the existence of the Atacama desert. Suggest why moister air currents do not affect the area from the east.

27 Study Table 12.5. Both Point Barrow and Sydney are coastal locations. Explain why their precipitation totals are so different.

Figure 12.27 Forests restrict air movement lowering wind speeds and preventing humid air from dispersing

Precipitation

It does not follow that a maritime location is always beneficial to the climate of an area, e.g. the existence of a cold offshore current will cool the air and lower temperatures. This may also encourage moisture to condense leading to coastal fog and mist (see section 8.3). By the time air currents reach the land, much moisture will have already been released and arid conditions often result. Generally, though, areas near a coast tend to have wetter climates. High levels of **evaporation** result in an increase in cloud cover, fog (see Fig. 8.19) and precipitation.

Table 12.5 Climate statistics for Point Barrow, USA and Sydney, Australia

Month	Jan	Feb	Mar	Apr	May	Jun	Jul	Aug	Sept	Oct	Nov	Dec
Point Barrow 71°N 156°E												
Temp (°C)	−26.4	−28	−26.1	−18	−7.5	2.2	8.4	6.1	−1.7	−8.3	−17.3	−23.6
Precip (mm)	5	3	3	3	3	8	23	20	13	13	8	5
Sydney 33°S 151°E												
Temp (°C)	21.95	21.95	20.8	18.05	15	12.5	11.7	13.05	15	17.5	19.45	21.1
Precipitation (mm)	89	102	127	135	127	117	117	76	74	71	74	74

12.5 The influence of other factors

Lakes, rivers, vegetation and soil also affect local climates. Temperatures near lakes are often lower than land in summer but higher in winter because of the differences in **thermal** properties. The moving water of rivers has more constant temperatures, though. Because of their smoothness, large areas of water encourage stronger and less turbulent winds than land areas, although humidities are higher.

The colour of vegetation and soils affect temperatures due to **albedo** (see section 2.4). In forest shades, temperatures are lower during the day but at night leaves trap radiant heat-keeping temperatures higher. Similarly, trees lower wind speeds while **evapotranspiration** from leaves increases humidity.

Summary

- Local climates or microclimates are those of a small area that possess clear contrasts with other nearby areas.
- Localised changes in the environment can lead to pronounced changes in atmospheric temperatures, wind speeds, humidity and precipitation. These in turn can affect the quality of the air allowing a build-up of pollutants that do not readily disperse.
- There are important links between air quality and health. An accumulation of pollutants can aggravate bronchial and respiratory complaints and is believed to be a contributory cause of asthma.
- One of the most distinctive and varied of all local climates is that of towns and cities. The built environment changes many aspects of the natural environment, particularly the roughness, thermal properties and hydrology. These combine to create an urban microclimate that can have important social, economic and ecological effects.
- Relief has an influence on winds, temperatures and precipitation. Temperature inversions can encourage smog and encourage hazardous conditions for people.
- Distance from the sea or continentality has an important influence on the weather and climate of an area. This is because the sea has a low thermal inertia while for land areas it is high.
- Lakes and rivers, vegetation and soils can also have important microclimatic effects.

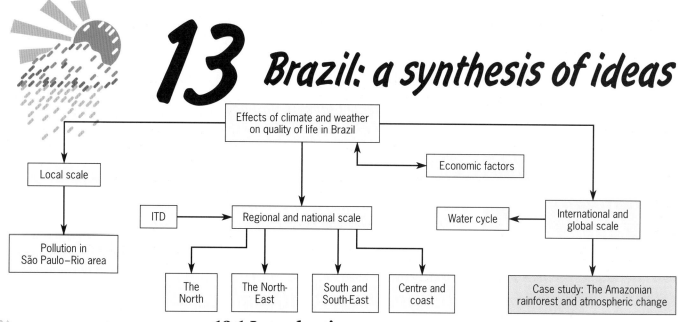

13 Brazil: a synthesis of ideas

Effects of climate and weather on quality of life in Brazil
→ Local scale → Pollution in São Paulo–Rio area
→ Economic factors
ITD → Regional and national scale → The North, The North-East, South and South-East, Centre and coast
International and global scale → Water cycle, Case study: The Amazonian rainforest and atmospheric change

13.1 Introduction

This chapter aims to bring together many of the concepts and issues raised throughout this book in the context of one country: Brazil. We will examine the hypothesis that the **climate** and **weather** in Brazil influence the quality of life of the people. We will also consider which issues relating to the climate in Brazil reflect the wider concerns we have studied in previous chapters. Brazil's environment has been the subject of attention from around the world, particularly during the Rio Summit in 1992, e.g. the destruction of the Amazonian rainforest, the largest single area of this ecosystem on earth. We will consider the extent to which human activity in the rainforest could influence weather and human activities elsewhere in Brazil and even the planet's climate. Such issues will be examined at a variety of scales, from the local to the regional and global.

Figure 13.1 Brazil: location in South America

Table 13.1 Socio-economic background of Brazil, 1990 (compared with the UK)

Area	GNP per capita (US$)	Share GDP accounted for by agriculture (%)	Per capita energy consumption (millions of tonnes of coal equivalent)	Infant mortality (%)	Life expectancy M	Life expectancy F	Urban population (%)
Brazil	2550	9	0.80	57	64	69	77
UK	14570	2	5.03	8	73	79	93

13.2 Brazil: an economically developing country

We will be looking at Brazil partly because of its size (Fig. 13.1). The country covers a large latitudinal and longitudinal range and encompasses many different climatic zones and natural regions.

Brazil is also one of the most industrialised countries in the economically developing world and the main industrial nation in South America. There are many signs of economic success in Brazil. These include some of the world's largest hydro-electric power schemes, modern industrial complexes and parts of cities, like São Paulo, that resemble those of more developed countries such as the USA. However, there are also severe economic problems and many parts of the country remain in a remote and very undeveloped state. In addition, the prosperity that has been achieved is not evenly distributed across the population and there is a large wealth gap within Brazil.

Brazil's urbanisation and industrialisation have also created some of the worst levels of industrial pollution in any area of the world. We will examine the nature of this pollution and assess its environmental impact.

?

1 Use Table 13.1 to make some comparisons between Brazil and the UK.

2 How do the problems faced by Brazil in Table 13.1 affect the country's ability to control environmental problems like pollution and damage to rainforests?

Figure 13.3 Pollution in the São Paulo–Rio industrial axis. Pollution here is so bad that people refer to parts of this area as 'the valley of death'.

Figure 13.2 The São Paulo–Rio de Janeiro industrial axis

13.3 Local scale: industrial pollution

Air quality and pollution in the São Paulo – Rio axis

Between Rio de Janeiro and São Paulo lies one of the economically developing world's most industrialised regions. The São Paulo – Rio axis (Fig. 13.2) stretches for about 430 km between the two cities and is home to over 700 000 companies. Many of these are heavy industries such as chemicals, oil refining, engineering, cars and pharmaceuticals. They located here during the rapid development that occurred in Brazil in the 'Golden years' of the 1970s.

With industrialisation came very high levels of atmospheric pollution that are typical of so many economically developing countries (Fig. 13.3). These gases represent a serious local disturbance to the atmospheric system that has many consequences for public health. The principal sources of this pollution are industry and transport, for environmental costs were largely ignored during the drive for development. In fact, the strict controls on emissions found in countries like the UK and the USA were also lacking for a long time. Pressure groups were (and still are) few in number and have limited influence (Fig 13.4). Finally, although a national environmental agency was established in 1973 it was largely ignored by the military government of the 1970s and early 1980s.

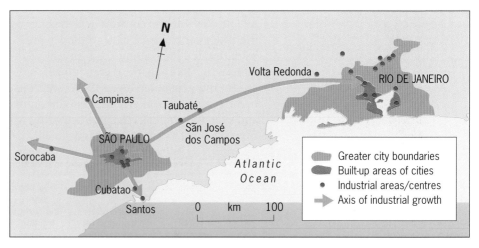

Since the mid-1980s, however, scientists became more concerned about the damage to the environment and undertook a number of research studies (Fig. 13.5). On the other hand, a huge amount of pressure has come from the international community to control damage to the rainforest and wildlife. Therefore, while scientists working in these fields have seen their budgets soar, '. . . the cities and industrial ghettos have remained poor relations, struggling for attention' (Joyce, 1985).

?

3 What do Figures 13.4–13.5 suggest are the main sources of pollution in Brazil's urban areas?

4 Study Figure 13.5. Outline some of the difficulties Brazil faces as it tries to control pollution and improve air quality with particular reference to: • economic development and • politics.

5 Outline some steps that could be taken to control levels of pollution in Brazil's cities.

TWENTY years ago Brazil welcomed polluters because polluters meant progress. Now it is probably the only country in the world to have pollution 'prosecutors' – a special team of lawyers who hunt down environmental offenders and take them to court.

As public prosecutors, selected by public examination and independent of government, they have an autonomy that allows them to take on both public and private companies. They have just won an out-of-court settlement from the French petro-chemical giant Rhône-Poulenc for 150 employees contaminated by toxic chemicals at their Cubatao factory.

They are suing one of Brazil's biggest private companies, CBA, for $10 million worth of environmental damage caused by a hydro-electric dam. 'But the government, at every level, federal state and municipal, is the biggest offender,' says Ronald Magri, the chief environmental prosecutor. The state energy company CESP has been taken to court many times, local authorities have been sued for polluting water sources and incinerating waste.

In São Paulo, Brazil's most advanced, industrialised and therefore polluted state, each of these 300 municipalities now has its special environmental prosecutor.

**Figure 13.4
Kicking up a stink in Rio**
(***Source:** Jan Rocha, The Guardian, 30 Nov. 1994*)

Figure 13.6 Brazil's major climatic
and natural regions

Scientists are beginning a long-overdue assessment of the damage already done. At the Catholic Pontifical University in Rio, for example, Dr Antonio Miguel has just completed studies of the combustion of gases produced by Brazil's millions of alcohol powered cars – something the government did not investigate when it decided to get off the petroleum treadmill. Miguel found a mixture of 85 per cent acid aldehydes and 15 per cent formal-dehyde... Formaldehyde is known to be a carcinogen in laboratory animals. But acid aldehydes are possibly a greater threat. They react in the atmosphere with nitrogen oxides and free radicals of oxygen to create PAN (peroxyacetylnitrate), a very strong mutagen, eye irritant and inhibitor of respiration.

Besides investigating combustion products from alcohol fuel, Miguel is sampling aerosols in Rio's air... Miguel is also trying to pinpoint the sources of pollutants by identifying the chemical signatures of various compounds that are in the air. For example, he looks for aluminium, iron and selenium that indicate dust...

Rio's second most serious problem (*after the lack of basic services*), said Mendosa (*former director of FEEMA**) is 'Pollution with the production of riches.' After Bhopal we are re-examining the chemical industry. The majority of these plants are in densely populated areas. Multinational companies have flocked to the São Paulo-Rio axis but, said Mendosa, most are more sensitive to public concern about pollution than their Brazilian counter-parts. Nonetheless, others at FEEMA recounted attempts by European and US companies to bluff the agency. 'The multinationals have the best consultants in the world' lamented one high-ranking official. With only a handful of second-hand instruments to measure pollutants, FEEMA must sometimes fall back on politics, he said. Not unlike environmental agencies in the US, FEEMA hopes that public pressure groups will complain loudly enough to give the agency political ammunition.

*FEEMA - Rio de Janeiro's state environmental agency

Figure 13.5 The price of progress in Brazil's cities (*Source: New Scientist*, 1985)

13.4 Regional and national scale: variety of climate

Table 13.2 is a compilation of climate statistics representative of Brazil's main climatic regions (Figure 13.6). The climate varies from the coast to the continental interior and although Brazil is usually thought of as a tropical country, there is even a more temperate climatic region in the south that occasionally suffers frosts. The climate and weather experienced throughout the year in different parts of Brazil are

Table 13.2 Average climate data for stations representing Brazil's main climatic regions.

		Jan	Feb	Mar	April	May	June	July	Aug	Sept	Oct	Nov	Dec
São Gabriel do Rio Negro alt. 85m	Temperature °C	25.4	25.6	25.5	25.2	25.0	24.5	24.3	24.8	25.4	25.6	25.9	25.4
	Precipitation (mm)	269	222	261	247	305	232	227	207	151	166	194	303
Manaus alt. 44m	Temperature °C	25.8	25.9	25.8	25.7	26.2	26.5	26.8	27.5	27.7	27.5	27.4	26.8
	Precipitation (mm)	277	251	318	286	221	94	52	46	64	111	163	217
Quixeramobim alt. 199m	Temperature °C	28.8	27.9	27.0	27.0	26.5	26.1	26.6	27.6	28.2	28.5	28.6	29.0
	Precipitation (mm)	42	121	209	173	118	58	16	5	6	3	4	10
Cuiabá alt.165m	Temperature °C	26.0	25.9	25.7	25.0	24.0	22.8	22.5	24.3	26.2	26.5	26.4	26.1
	Precipitation (mm)	214	196	231	102	49	20	3	14	40	131	140	219
São Paulo alt. 796m	Temperature °C	20.8	20.9	20.6	18.8	16.6	15.4	14.9	15.5	16.3	17.5	18.5	20.1
	Precipitation (mm)	215	175	161	77	65	40	24	48	92	121	138	188
Pôrto Alegre alt. 15m	Temperature °C	24.5	24.7	22.8	22.7	17.2	13.5	13.7	14.6	16.5	18.3	21.3	22.8
	Precipitation (mm)	109	93	91	167	104	127	109	129	116	78	83	104
Curitiba alt. 907m	Temperature °C	19.8	19.8	19.6	16.6	14.6	13.4	12.9	13.5	14.4	16.1	17.4	19.2
	Precipitation (mm)	179	152	102	92	80	76	60	92	125	144	131	146
Salvador alt. 47m	Temperature °C	27.0	27.0	27.0	26.0	26.0	25.0	24.0	24.0	26.0	27.0	27.0	27.0
	Precipitation (mm)	66	130	154	284	274	239	183	122	83	102	114	142

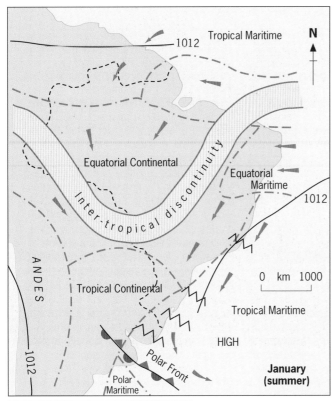

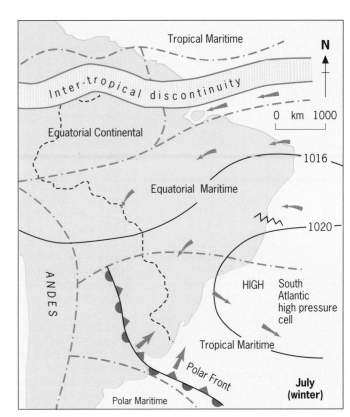

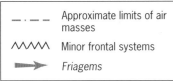

- - - - · Approximate limits of air masses

∧∧∧∧ Minor frontal systems

➤➤➤ *Friagems*

Figure 13.7 Seasonal changes in air masses, winds and pressure systems affecting Brazil

6a Use the climate statistics in Table 13.2 to draw climate graphs for Manaus, Cuiabá, Quixeramobim, Salvador and Curitiba to represent each main climate region shown in Figure 13.6.
b With reference to their location in Brazil, describe and account for the differences between the climates of the following locations. Use Figures 13.7 and 13.8 to help you.
• Manaus and São Gabriel do Rio Negro
• Curitiba and Pôrto Alegre

7 Study Table 13.2 and Figure 13.6. Discuss how you would expect climatic variations in Brazil to influence human activities. Use the economic maps of Brazil in an atlas to help you.

also strongly influenced by the changing position of the **Inter-Tropical Discontinuity** (**ITD**) (Figs 13.7 and 13.8) and the way in which this affects the winds and air masses over the country.

The North

Northern Brazil experiences a hot, wet climate (Figs 13.6 and 13.9). Although the temperatures can be accounted for relatively easily, an explanation of the rainfall is less clear. In addition, there is a lack of data, as records of rainfall are sparse over Amazonia. We have previously established that the ITD is not a uniform zone of convergence and storms (see section 10.2). Meterologists have not thoroughly studied storm lines associated with the ITD in Brazil, but it is likely that strong local turbulence and **convection** (see section 2.5) produce a less regular distribution of storms. **Humidities** can be as high as 70 per cent during the wet season and in some places nearer 80 per cent.

Shower activity seems to follow a spatial pattern, with March being one of the wettest months in Amazonia. A comparison of three stations in the west (Uapes), centre (Manaus) and east (Santarem) of Amazonia has shown that in the west rainfall is concentrated over 12 days, in the centre over 20 days and in the east over 26 days (Boucher, 1974). This suggests that storms are heavier but occur less regularly further south and east. It is likely that surges in the north-west air flow account for much of the rainfall and that these are stronger further east. After March, the North-Westerlies are replaced again by the drier South-Easterly Trades marking the return of the drier period of the year.

The natural vegetation over much of northern Brazil is dense tropical rainforest (see Fig. 13.15). It is a major source of moisture through **evapotranspiration**.

The North-East

This region (Fig. 13.6) is something of a puzzle in Brazil. It is a semi-arid region, yet it is on a similar latitude to the wetter north region and is near the coast, protruding into the Atlantic.

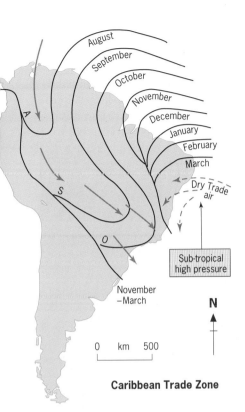

Figure 13.8 Changing position of the ITD and north-westerly air flow (*Source:* Boucher, 1974)

Table 13.3 Migration between Brazil's regions

Region	Immigration 1970	Immigration 1980	Emigration 1970	Emigration 1980
North	9.90	18.16	7.21	7.35
North-East	5.93	6.49	19.26	19.46
South-East	16.57	18.45	15.60	13.61
South	17.50	14.06	10.18	14.47
Central-West	32.84	35.14	8.35	13.20

8 With reference to Table 13.2 and Figure 13.7 describe and account for the climate of Quixeramobim.

9 Study Figures 13.11 and Table 13.3. Comment on the problems that result from the severe climate of the drought polygon under the following headings: • social • economic • political.

10 Look at Figures 13.9 and 13.10. Compare how the climate has influenced housing and lifestyles in each region.

Figure 13.9 Housing in Brazil's North region

Figure 13.10 Housing in Brazil's North-East region

The impact of the drought polygon

The arid north-east corner of Brazil is known as the *Poligono das Secas*, or drought polygon. It is a semi-arid region (Fig. 13.10) of sparse, largely thorn scrub vegetation and many of the plants, including cacti, are *xerophytic* (adapted to shortages of water). The core of the region has annual rainfall of less than 500 mm, although this is unreliable and is often less in years of drought. When the rains do come, they occur as intense showers that lead to flash floods. Consequently, much of the precious moisture is evaporated or lost in runoff rather than infiltrating the soil. Inevitably, this leads to severe hardship for the poorer people living in North-East Brazil (Fig. 13.11). In fact, 1993 was the fourth consecutive year of lower than normal rainfall in the region.

Drought victims loot Brazil's food shops

Starving peasants have invaded many towns in the drought-ridden interior of North-East Brazil to plunder government warehouses and supermarkets. About nine million people are facing great hardship.

In Serra Talhada, in the state of Pernambuco, hundreds of peasants swarmed into the football stadium on hearing that 30 tonnes of beans were to be distributed to drought victims.

After waiting patiently in the hot sun for four hours, many of them, gaunt, dressed in rags and carrying children, marched to the large street market in the centre of town.

'They were grey with hunger,' one of the stall-holders said. 'Some stuffed handfuls of flour into their mouths. Others ate bars of raw sugar, some cramming bananas, skin and all, into their mouths.'

Last week, 500 peasants travelled to Recife, the capital of Pernambuco. With the help of Brazil's main rural trade union, they occupied the headquarters of Sudene, the federal government's development agency for the North-East. After holding the super-intendent of Sudene and several local politicians hostage for nine hours, the peasants won the promise of a meeting with President Itamar Franco.

Held the following day, the president promised to spend $180 million in emergency food aid. Mr Franco had been reluctant to take this step becuase of the corrupt distribution network, dominated by old-style political bosses.

Despite the president's promise, the plundering has increased. According to peasant folklore, the North-East could expect a prolonged drought if heavy rain did not fall on March 19, the feast of St Joseph of Nazareth. It did not, and many families are convinced they face a year of destitution.

Meteorologists are also gloomy. ' We expect it to rain more this year than last. But as this is the fourth year of drought, the social impact will be far greater,' Carlos Nobre, director of the region's meteorological centre, said.

Figure 13.11 The impact of drought in the North-East region of Brazil (*Source:* Sue Branford, *The Guardian*, 25 March 1993)

?

11 With reference to Table 13.2 and Figure 13.7 describe and account for the differences between the climate of Salvador and Cuiabá which are on approximately the same latitude.

12 Account for the changing patterns of land use along the transect in Figure 13.13.

13a Construct a line graph based on the data in Table 13.4.
b Add a curve for the running mean from 1970 to 1988.
c Describe and explain the trend in total production.

14 Brazil has a well-balanced economy and unlike many poorer countries in the world it does not rely too heavily on the export of one or two primary products. Coffee once accounted for over 50 per cent of Brazil's export earnings but today it accounts for about 4 per cent (1993). Comment on the economic consequences of such fluctuations in coffee production.

Table 13.4 Brazil's total coffee production, 1970–80 (*Source:* International Coffee Organisation)

Year	'000s of bags
1970	19 719
1971	28 098
1972	21 683
1973	21 600
1974	26 702
1975	14 544
1976	11 355
1977	18 451
1978	21 074
1979	19 302
1980	23 946
1981	26 665
1982	23 410
1983	24 742
1984	25 873
1985	22 808
1986	29 160
1987	31 822
1988	25 300

The centre and coastal areas

The centre of Brazil is an area of plateau with grassland vegetation. Although the climate would encourage dense vegetation, the soils are highly weathered and acidic. This is not suitable for arable farming but does attract herds of grazing animals (Fig. 13.12). The land use of the central west region is therefore dominated by pastoral farming. Brazil has over 100 million beef cattle which is the largest herd in South America and the fourth largest in the world. Towards the coast, agriculture becomes more diverse and there is a mixture of both arable and pastoral activities.

Figure 13.12 Cattle farming in Brazil's Central region

The South and South-East

The South experiences a more sub-tropical to temperate climate, in part depending on altitude. One particular synoptic feature deserves explanation. About two or three times a year during the 'winter' months, waves of cold air migrate away from the polar **front**. Very occasionally these *friagems* extend as far north as parts of Amazonia, bringing sudden and unexpected cold air (see Fig. 13.7). On one such occasion in 1963, the Amazonian Indians experienced considerable discomfort as they were unaccustomed to such cold. In eastern Brazil these waves of cold air pass along the coast at times when the sub tropical high pressure (STHP) cell has weakened and moved eastwards, this allows the colder air to penetrate further north.

Friagems can bring frosts to the South and South-East regions of Brazil. Such frosts are a major hazard particularly to farmers in southern Brazil. There have been many occasions, as in 1963, 1969/70 and 1972/73, when frosts seriously affected the coffee harvests with devastating consequences. The worst year in recent history was 1975 when severe frost damage reduced coffee production by 75 per cent. As a result, some African coffee had to be imported and then exported by Brazil to satisfy the domestic markets.

Figure 13.13 Diagrammatic section showing main agricultural land uses and relief in Brazil (along latitude 20°S approximately)

| | Rice | Commercial beef farming | Sugar, coffee, tomatoes, oranges, pineapples, some pastoral | Bananas, fruit |

Commercial beef farming

2000 m (asl approx.)
Matto Grosso
1000

Brazilian Highlands

Atlantic Ocean

Sea level

13.5 International and global scale: climate change

Amazonian rain forest and the water cycle

Figure 13.14 summarises the findings of Dr Eneas Salati, a former director of CENA (Brazil's Centre for Nuclear Energy and Agriculture). Salati believes that the climatic equilibrium that exists in northern Brazil has evolved step by step with the development of the Amazon rainforest. Both are equally dependent on each other. There is an inescapable conclusion that destroying large areas of rainforest will alter the water cycle in this part of South America and so influence patterns of rainfall and human activities over a much wider area.

Figure 13.14 The water cycle over the Amazonian rainforest and its effects on other parts of Brazil (*After:* Salati, 1969)

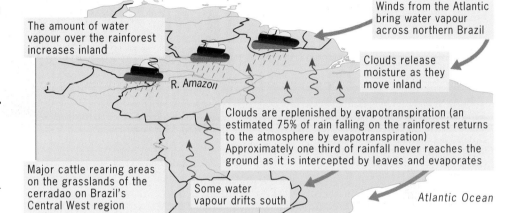

The amount of water vapour over the rainforest increases inland

Winds from the Atlantic bring water vapour across northern Brazil

Clouds release moisture as they move inland

R. Amazon

Clouds are replenished by evapotranspiration (an estimated 75% of rain falling on the rainforest returns to the atmosphere by evapotranspiration) Approximately one third of rainfall never reaches the ground as it is intercepted by leaves and evaporates

Major cattle rearing areas on the grasslands of the cerradao on Brazil's Central West region

Some water vapour drifts south

Atlantic Ocean

?

15 Use Figure 13.14 to explain the links between the Amazonian rainforest and the weather over this part of South America.

The Amazonian rainforest and atmospheric change

We frequently call the Amazon rainforest 'the lungs of the earth' (Fig. 13.15). In fact, research since 1985 suggests that the rainforest in Brazil may be a key element in the workings of the earth's atmosphere. Changes in Amazonia in the pattern of human activities could therefore have a dramatic impact on the planet's climate. For example, scientists have questioned the contribution of rainforests to the atmosphere's budget of trace gases, and most of us have asked what will happen to the atmosphere as deforestation continues.

Trace gases

Carbon dioxide (CO_2) has received much attention as a greenhouse gas (see section 4.3). It absorbs radiant heat from the earth's surface and traps it in the **troposphere** leading to a warming of the world's climate. However, minor trace gases are arguably just as important in affecting climate change and rainforests may be the most important sources of these gases to the atmosphere. We have seen that gases like nitrous oxide (N_2O) and methane (CH_4) trap infrared radiation from the earth. These gases in turn also affect the amount of ozone in the **stratosphere** (see section 5. 2) that protects organisms form the harmful effects of ultraviolet radiation from the sun. Carbon monoxide (CO), another trace gas, destroys hydroxyl radicals (a chemical compound containing hydrogen and oxygen) which cleanses the atmosphere of methane.

During the 1970s scientists observed that levels of these gases were rising. This was largely blamed on industrialisation, high altitude aircraft and changes in agricultural practices (see section 4.3). Although some scientists felt that 'we were sort of stabbing in the dark' for explanations (Tom Goreau, University of Miami), all of these activities were shown to be important. However, changes in the oceans, farming and industrialisation did not account for the world's changing balance of CO, CO_2, N_2O and CH_4.

Figure 13.15 Aerial view of dense Amazon rainforest, Brazil

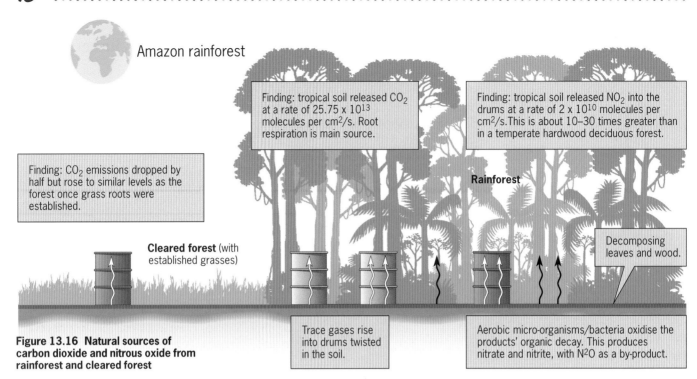

Amazon rainforest

Finding: CO_2 emissions dropped by half but rose to similar levels as the forest once grass roots were established.

Finding: tropical soil released CO_2 at a rate of 25.75×10^{13} molecules per cm^2/s. Root respiration is main source.

Finding: tropical soil released NO_2 into the drums at a rate of 2×10^{10} molecules per cm^2/s. This is about 10–30 times greater than in a temperate hardwood deciduous forest.

Rainforest

Cleared forest (with established grasses)

Decomposing leaves and wood.

Figure 13.16 Natural sources of carbon dioxide and nitrous oxide from rainforest and cleared forest

Trace gases rise into drums twisted in the soil.

Aerobic micro-organisms/bacteria oxidise the products' organic decay. This produces nitrate and nitrite, with N^2O as a by-product.

During the 1980s scientists turned their attention to the rainforests and carried out extensive measurements of gases, especially in Amazonia. Here, after cutting part of the forest, farmers usually burn the vegetation and the plumes from these fires contain several gases.

Levels of carbon monoxide and carbon dioxide

Aircraft flying over Amazonia measured CO in quantities higher than expected. This may be because less oxygen is available at ground level to fuel the fires than was once thought. The fires also produce huge amounts of CO_2, possibly adding as much to the atmosphere as the worldwide burning of fossil fuels. These fires also release some N_2O. However, there was a problem. Forest fires alone did not account for the 0.2 per cent a year build-up in N_2O or the 1–2 per cent increase in CH_4. So it was thought that natural sources may be more important, leading to further research in the mid-1980s (Fig. 13.16).

Researchers concluded that mature grassland cycles nutrients very quickly and so carbon is not stored for long periods as it is in forested areas. It was therefore feared that the cutting down of large areas of rainforest would reduce their ability to store large amounts of CO_2 and thus alter levels of atmospheric CO_2 as seriously as the burning of fossil fuels. We have already established that this could lead to higher world temperatures with many other effects on the planet (see section 4.4).

Levels of nitrous oxide

A number of studies have revealed that tropical rainforests may generate up to a half of the 15 million tonnes of atmospheric nitrous oxide that the planet produces each year. Only a third results from human activities like burning vegetation or from fertilising crops. None the less, there has been an increase in N_2O of about 10–20 per cent on the pre-industrial world levels. This is a matter for concern, as much N_2O rises into the stratosphere where it becomes nitric oxide that damages ozone. Nitrous oxide is also another greenhouse gas. Deforestation could therefore benefit the **global warming** issue as it would slow the build-up of N_2O in the atmosphere.

Levels of methane

Rainforests are a source of methane which bubbles up from the flooded wetlands, called *varzea*, as a result of the decomposition of organic matter. Insects like termites also produce methane as a waste product as they digest vegetation. However, tropical rainforests are home to bacteria in the soil that consume methane. Consequently, what remains an area for research is whether the Amazon is a net source or *sink*.

In conclusion, there is no doubt that tropical rainforests like Amazonia make a major contribution to the balance of gases in the earth's atmosphere. The destruction of rainforests could therefore have long-lasting effects on the world's climate that we cannot fully predict or understand at present.

?

16 Explain how human activities in other parts of Brazil and South America could be affected by climatic changes brought about by destruction of Amazonian rainforest.

17 Essay: Assess the importance of tropical rainforests to the world's climate.

?

18   To what extent do you think the hypothesis that climate and weather have a major effect on people's quality of life has been proven in this chapter?

19a   Assess the relevance of the five strategic goals in Figure 13.17 to the major problems of pollution and deforestation and their effect on the atmosphere that we have studied in Brazil.

b   Explain how each of the suggestions in Figure 13.17 could help to combat global climate change.

c   Compile a similar list of steps that could be taken by individuals to help accomplish the five goals.

20   Essay: 'We will not protect the environment until we address the issue of poverty and population growth in the same breath' (Prince Charles, date). Assess the relevance of this statement to the problem of climate change resulting from human activities.

13.6 Conclusions

We have seen that Brazil faces many pressing economic and social problems. As a country it has received much criticism from the international community, particularly from the economically developed world, for its environmental record. However, not only do developed countries have an equally poor record of environmental damage but they also have far greater available resources to tackle pollution control and repair environmental mismanagement.

Since the 19th century and particularly in the latter half of the twentieth century, human activities have increasingly had an effect on the atmosphere. In this book we have studied issues such as **acid rain** and other types of pollution, deforestation and dramatic changes from rural to urban landscapes in many parts of the world. Climate change may even pose a threat to our security in the future. We have seen in Brazil that tampering with atmospheric processes may alter the water cycle and spread the influence of drought to other parts of the world. Future world conflicts may therefore not be over oil or minerals but over productive land and water supplies. For example, tensions could easily arise if large numbers of people become refugees as a result of coastal flooding linked to global warming.

We can usefully, though, consider these issues in a broader context. They are the result of larger scale human changes and in particular population growth, economic development and increasing levels of resource use. It is vitally important that we see the system as a whole and recognise that concern over climate change, destruction of the **ozone layer** and global warming is of little use if other pressing issues are not tackled at the same time.

Figure 13.17 Vice-President Al Gore's five strategic goals for saving the world's environment (*Source:* Houghton, 1990)

Al Gore, the vice-president of the United States, has proposed a plan for saving the world's environment. He has called it 'A Global Marshall Plan' paralleled after the Marshall plan through which the United States assisted western Europe to recover and rebuild after World War 2. Resources for the plan would need to come from the world's major wealthy countries. He has proposed five strategic goals for the plan:
1   The stabilisation of world population.
2   The rapid creation and development of environmentally appropriate technologies.
3   A comprehensive and ubiquitous change in the economic 'rules of the road' by which we measure the impact of our decisions on the environment.
4   The negotiation and approval of a new generation of international agreements which must be especially sensitive to the vast differences of capability and need between developed and developing nations.
5   The establishment of a co-operative plan for educating the world's citizens about our global environment.

Summary

- As Brazil covers about one third of the area of South America and spans such a large latitudinal and longitudinal range, it encompasses many different climatic zones.
- The São Paulo – Rio industrial axis is one of the most heavily polluted areas in the world. The atmospheric pollution and poor air quality create a health hazard for Brazilians living in these cities.
- Brazil's climatic regions range from the humid equatorial areas of the North and the arid North-East to the continental interior climate of the Central-West, the milder and wetter coastal areas and the sub-tropical South and South-East.
- Brazil's climate and weather have an important effect on people's quality of life. Extremes such as drought and sudden frosts have even influenced the stability of Brazil's economy.
- The destruction of large areas of rainforest will alter the water cycle in this part of South America and influence patterns of rainfall and human activities over a much wider area.
- Research suggests that the rainforest in Brazil may be a key element in the workings of the earth's atmosphere. Changes there in the pattern of human activities could have a dramatic impact on the planet's climate.
- Environmental issues relating to atmospheric processes are the result of larger scale human changes.

References

Acid Rain Information Centre (1987) *Acid Deposition in the Northern Hemisphere,* a publication for the European Year of the Environment.

Allayne Street-Perrott, F (1994) 'Drowned trees record dry spells', *Nature,* vol. 369, p.518.

Alley, R B, Meese, D A (1993) 'Abrupt increase in Greenland snow accumulation at the end of the Younger Dryas event', *Nature,* vol. 362, 8 April 1993, p. 527

Auer, A H (1978) 'Correlation of land use and cover with meteorological anomalies', *Journal of Applied Meteorology,* vol. 17, No. 5 May, pp.636–43.

Australian Bureau of Statistics (1992) *Australia's environment: issues and facts.*

Australian Bureau of Statistics (1993) *Environmental issues: people's views and practices.*

Ayoade, J (1990) 'Climatic aspects of solar and wind energy in Nigeria', *Malaysian Journal of Tropical Geography,* June vol. 21, No. 1, pp.1–8.

Barry, R G (1981) *Climate change,* Macmillan Education.

Barry, R G (1992) *Mountain weather and climate,* Routledge.

Barry, R G and Chorley, R J (1987) *Atmosphere, weather and climate,* Routledge.

Baker, E J and Patton, D J (1974) 'Attitudes towards hurricane hazards on the Gulf Coast', White, G F, *Natural Hazards,* p.30–3, OUP.

Baumann, D D and Sims, J H (1974) 'Human response to the Hurricane', White, G F, *Natural Hazards,* p.25–30, OUP.

Boucher, K (1974) *Global Climate,* English Universities Press Ltd.

British Overseas Development, report, May 1993 No. 28.

Carpenter, C (1991) *The changing world of weather,* Guinness Publishing Ltd.

Chandler, T J (1972) *Modern Meteorology and Climatology,* Nelson.

Changnon, S A (1978) 'Urban effects on severe local storms at St Louis', *Journal of Applied Meteorology,* vol. 17, No. 5 May, pp.578–86.

Department of the Environment (1991) *Global Atmosphere and Air Quality*

Department of the Environment (1991) *The Ozone Layer.*

Fairbanks, R G (1993) 'Flip-flop end to last ice age', *Nature,* Vol. 362, p.495.

Gregory, S (1993) 'A hundred years of temperature and precipitation fluctuations at Sheffield 1891 to 1991' *Geography,* Vol. 78, pp.241–249, The Geographical Association.

Gribbin, J (ed.) (1986) *The breathing planet,* Basil Blackwell and New Scientist.

Gribbin, J (1988) *The oceanic key to climatic change,* New Scientist, May p.32.

Gribbin, J (1991) *Climate now,* New Scientist Inside Science No. 44.

Hilton, K (1979) *Process and Pattern in Physical Geography,* Bell and Hyman.

HMSO (1994) *Climate Change: The UK programme.* Crown copyright is reproduced with the permission of the Controller of HMSO.

Houghton, J T (ed.) (1990) *Climate change, the IPCC scientific assessment,* WHO and UNEP (CUP).

Houghton, J T (1994) *Global Warming,* Lion Publishing.

Huff, F A and Vogel, J L (1978) 'Urban, topographic and diurnal effects on rainfall in the St Louis region' *Journal of Applied Meteorology,* vol. 17, no. 5, May pp.565–77.

Jackson, I J (1989 second ed.) *Climate, water and agriculture in the Tropics,* Longman.

Johnson, B L C (1979) *India,* Heinemann Educational Books.

Joyce, C (1985) 'The price of progress', New Scientist, July pp.46–50.

Lamb, H H (1982) *Climate, history and the modern world,* Methuen.

Legates, D R and Mather, J R (1992) 'An evaluation of the average annual global water balance', *Geographical Review,* vol. 82, no, pp.253–67.

Leggett, J (1994) *Who will underwrite the Hurricane?,* New Scientist, 7 August pp.29–31.

Leggett, J (1994) *Climate Change: How bad is it? What is being done? What needs to be done?* Reinsurance Association of America, 29 April.

Lockwood, J G (1976) *World Climatology,* Edward Arnold.

Manley, G (1970) 'Climate in Britain over 10000 years, *The Geographical Magazine,* Vol. 43, no. 2.

Meteorological Office *Climate of the British Isles,* Leaflet ES4.

Meteorological Office *Water in the atmosphere,* Leaflet ES6.

Meteorological Office *Weather Satellites,* Leaflet ES8.

Meteorological Office *Weather Forecasting,* Leaflet ES11.

Meteorological Office and Department of the Environment (1989) *Global Climate Change.* HMSO. Crown copyright is reproduced with the permission of the Controller of HMSO.

Morgan, W T W (1973) *East Africa,* Longman.

Oeschger, H and Stauffer, B (1986) 'Review of the history of atmospheric CO_2 recorded in ice cores', *The Changing Carbon Cycle: A Global Analysis.* Ed Trabalka, J R and Reichler, D E, pp.89–108, Springer Verlag.

Olstead, J (1993) 'Global warming in the dock', in *The Geographical Magazine,* September pp.12–16.

Ozone Science Unit (1992) *Depletion of the Ozone Layer,* Australian Bureau of Meteorology.

Parry, M (1990) *Climate Change and World Agriculture,* Earthscan Publications Ltd.

Paterson, D (1993) *Did Tibet cool the world?,* New Scientist, July pp.29–33.

Pearce, E A and Smith, C G (1993) *The World Weather Guide,* Helicon.

Pearce, F (1993) *El Niño roars back against the odds,* New Scientist, May pp.7.

Ramachandran, R and Thakur, S C (1974) 'India and the Ganga floodplains', White, G F *Natural Hazards,* pp.43–46, OUP.

Rosenweig, C (1994) 'Maize suffers a sea-change', in *Nature,* vol. 370, pp.175–6.

Subbaramayya, I, Babu, S V and Rao, S S (1984) 'Onset of the summer monsoon over India and its variability' in *Meteorological Magazine,* vol. 113, pp. 127–135.

Tapper, N J, Tyson, P D, Owens, I F and Hastie, W J (1981) 'Modelling the winter urban heat island over Christchurch, New Zealand', *Journal of Applied Meteorology,* vol. 20, no. 4, April pp.365–376.

Glossary

Acid rain   All rainfall is naturally slightly acidic. This term generally refers to the increase in acidity of **precipitation** due to pollution.

Advection   The movement of air and moisture in a horizontal direction.

Advection fog   Fog which results from air passing horizontally over a cold surface that encourages **condensation.**

Air mass   A body of air with more or less uniform characteristics, particularly temperature and **humidity.**

Albedo   This refers to the reflectivity of a surface. It is the ratio between the amount of incoming **insolation** and the amount of reflected energy expressed as a decimal or a percentage.

Ana front   The front in a depression which occurs where air is generally rising. These unstable conditions encourage cloud development and heavy **precipiation.**

Anabatic winds   Light winds rising up valley slopes in mountain areas during the day.

Anticyclone   A high **pressure** system with closed **isobars** and diverging air near the ground. Winds in the northern hemisphere move clockwise in the system.

Aspect   The direction in which a slope faces.

Blocking anticyclone   Large anticyclones, or highs, that 'block' or deflect **depressions** moving over an area.

Boundary or friction layer   The lowest 1000m or so of the **troposphere** in which surface winds occur, e.g. land and sea breezes.

Chaos theory   A theory that suggests that although natural processes (like those in the atmosphere) are largely stable, unpredictable, random changes can suddenly occur.

Circumpolar vortex   A powerful flow of air currents in the upper **troposphere** that travels around the North Pole. These air currents occur in the mid-latitutudes and are also called the **Ferrel westerlies.** See also **polar vortex.**

Climate   The average **weather** over an area. Larger temporal and spatial scales are involved when describing the climate of an area, although the same descriptive elements are used.

Cloud seeding   The artificial process of encouraging **precipitation** by the introduction of freezing, **condensation nuclei** into clouds.

Condensation   The process that involves water changing from a gas to a liquid.

Condensation point or level   The point (temperature) or altitude at which the **dew point** temperature is reached and **condensation** occurs.

Condensation nuclei   Small particles in the atmosphere that attract water and encourage **condensation** (see also **hygroscopic particles**).

Continentality   The influence of inland locations, as opposed to locations near the sea, on **weather** and **climate.**

Convection   The upward movement of air as a result of heating from contact with the ground or sea. Air also moves in a downward motion within a convection cell.

Coriolis force   An imaginary force resulting from the rotation of the earth. It causes winds to deflect to the right in the northern hemisphere and to the left south of the equator.

Cyclone   A low **pressure** system with closed **isobars** and converging air at the ground. In the northern hemisphere winds travel anticlockwise in the system and include **depressions** and **hurricanes.**

Depression   A low-intensity **cyclone.**

Dew point   The temperature at which a body of air becomes saturated.

Diurnal   Daily e.g. daily variations in temperature.

Doldrums   An area of very gentle, variable winds occasionally broken by thunderstorms. It is associated with the low **pressure** which occurs on the equatorial limb of the **Hadley cell.**

El Niño Southern Oscillations   Periodic reversals in the normal westward flow of the trade winds and ocean currents that flow across the tropical Pacific from the Americas towards Asia. ENSOs disrupt normal climatic patterns across much of the world, including the **monsoon** in south Asia.

Elements (of weather and climate)   Various aspects of the atmosphere which are used to describe **weather** and **climate** e.g. **precipitation, pressure,** wind, cloud cover, **humidity,** sunshine and temperature.

Equinox   When the overhead sun is directly above the equator at noon, resulting from the earth's axis being at a tangent to the sun on 21 March and 22 September each year.

Evaporation   The process that involves water changing from a liquid into a gas; water vapour.

Evapotranspiration   A combination of transpiration and **evaporation**. Separating these processes for measurement is very difficult, so they are usually combined.

Ferrel westerlies   A belt of high altitude westerly winds which occur in the middle latitudes. They are locally concentrated near the top of the **troposphere** into **jet streams**.

Fohn   A type of wind which blows in mountain areas.

Front   This is a zone or boundary in the atmosphere separating two **air masses**. It is represented symbolically on weather maps.

Geostrophic wind   An air current that moves parallel to **isobars** as a result of the **pressure gradient** force and the **Coriolis force** being in balance.

Gradient wind   An air current moving parallel to curved **isobars** as a result of a balance between the **Coriolis** and centrifugal forces.

Greenhouse effect   The natural warming of the atmosphere which occurs when certain gases in the atmosphere, like carbon dioxide, absorb outgoing long-wave radiation from the earth's surface.

Global warming   A process similar to the **greenhouse effect**, but this is the artificial warming of the earth's atmosphere as a result of pollution.

Gyre   A circulation of water which occurs in each of the major ocean basins between about 20° and 30° north and south of the equator.

Hadley cell   A circulatory cell between the equator and the tropics named after the English astronomer E. Hadley.

Heat island   A zone of higher air temperatures associated with urban areas. Heat islands are created by the release of waste heat e.g. domestic, commercial and industrial sources.

Humidity   A measure of the concentration of water vapour in a body of air.

Hurricane   An intense tropical **cyclone** accompanied by strong winds (over 160km/h) and heavy rainfall.

Hydrological cycle   The cycle of moisture which occurs within and between the earth and its atmosphere.

Hygroscopic particles   Particles like dust and salt in the atmosphere that attract water (see **condensation nuclei**).

Insolation   An alternative way of referring to short-wave radiant energy from the sun.

Inter-Tropical Convergence Zone   A virtually continuous belt of low **pressure** near the equator where the north-easterly and south-easterly trade winds converge.

Instability   A condition of air which leads to air rising because the environmental **lapse rate** is greater than the adiabatic lapse rate.

Inter-Tropical Discontinuity   Another name for the **Inter-Tropical Convergence Zone**. A term that more accurately reflects the discontinuous nature of this zone.

Ionosphere   The outer layer of the atmosphere, synonymous with the **thermosphere**. In this zone atmospheric gases are ionised by incoming **insolation** and temperatures rise with height.

Isobar   A line on a weather map joining points of equal atmospheric **pressure** – measured in millibars.

Isohyet   A line on a weather map joining points of equal rainfall – measured in millimetres.

Jet stream   A narrow belt of very fast moving air near the top of the **troposphere**.

Kata front   The front in a **depression** associated with generally descending air. Cloud development is restricted and there is only light rain.

Katabatic winds   Light winds moving down valley slopes in mountain areas at night.

Lapse rate   The rate at which temperature changes with altitude. This is usually inverse; temperature decreasing with altitude in the **troposphere**.

Latent heat   The amount of heat energy needed to convert water into water vapour. It is energy that is also released when water vapour **condenses.**

Mesosphere   A zone between about 50 and 90 km above the earth's surface.

Mesopause   The boundary between the **mesosphere** and the **ionosphere**.

Meteorology   The scientific study of **weather**, **climate** and the earth's atmosphere.

Meteorological equator   An area of **low pressure** also referred to as the **Inter-Tropical Discontinuity** or **Inter-Tropical Convergence Zone**. Unlike the 0° line of latitude, the meteorological equator moves with the changing relative position of the overhead sun.

Microclimate   The climate experienced over a very small area e.g. a lake, urban area or even surrounding a leaf.

Milankovitch cycles   Three ways in which the earth's geometry varies as calculated by Milutin Milankovitch in the 1920s.

Monsoon   The seasonal reversal of **pressure** systems and winds over land areas and neighbouring oceans. It more popularly also refers to a period of intense rainfall associated with these atmospheric changes.

Occlusion   When the warm air in a **depression** is lifted above the ground and the cold air behind the cold **front** meets the cold air ahead of the warm front. There is often vertical motion of air, cloud development and rain.

Orographic   This refers to the influence of relief e.g. on **precipitation**.

Ozone layer   A layer of the **stratosphere** of concentrated ozone (O_3) molecules. It particularly filters ultra-violet short-wave radiation from the sun.

Polar vortex   A belt of strong westerly winds occurring in the winter around Antarctica. The polar vortex separates the polar **stratosphere** from the tropical stratosphere.

Potential evapotranspiration   The maximum possible amount of **evapotranspiration** at a given temperature, given an unlimited supply of water.

Precipitation   When moisture falls from the atmosphere. This can occur in a variety of forms e.g. rain, hail, sleet and snow.

Pressure   The weight of the atmosphere or a column of air above a particular point. Atmospheric pressure is the result of gravity and decreases with height.

Pressure gradient   The gradient in **pressure** between an area of low and an area of high atmospheric pressure. This is reflected in the spacing of **isobars** on a **synoptic weather chart**.

Prevailing wind   The most frequently occurring direction from which a wind blows.

Radiation fog   Fog that results from the **condensation** of moisture when moist air is cooled by the ground as it sinks into valleys at night.

Rain shadow   An area on the leeward side of mountains that receives less **precipitation** than on the other side. This is because prevailing winds release much of their moisture over the mountains. By the time air reaches the leeward side, or rain shadow, it is comparatively dry. Note that the amounts of precipitation are relative. A rainshadow area can still experience a wet climate.

Relative humidity   The ratio between the amount of water vapour held in a mass of air at a given temperature, and the total amount (100 per cent) of water vapour that could be held at that same temperature, expressed as a percentage.

Ridge   An elongated area of high pressure.

Rossby waves   A series of large waves that occur in the westerlies in the middle latitudes in each hemisphere.

Sea surface temperature anomalies   Irregular, large scale variations in sea surface temperatures that scientists believe are an important influence on climate in different parts of the world.

Smog   A combination of fog and pollution. Pollutants are often **hygroscopic particles** that act as **condensation nuclei**.

Solar budget   The balance between incoming solar energy and outgoing radiant energy.

Solar constant   The virtually constant amount of solar energy received at the outer edge of the earth's atmosphere, represented as $1350Wm^{-2}$.

Solstice   When the overhead midday sun is at its greatest angular distance from the equator on 21 June and 22 December. The overhead sun's position is approximately $23\frac{1}{2}°$ north and south over the tropics.

Storm surge   A rise in the level of the sea resulting from an area of low **pressure**, like a **hurricane**. This leads to flooding of low-lying coastal areas.

Stability   The tendency of air not to be buoyant when its temperature is cooler than the surrounding environmental air. The opposite is **instability**.

Stratopause   The boundary in the atmosphere between the **stratosphere** and the **troposphere**.

Stratosphere   The layer in the atmosphere above the **troposphere**.

Sublimation   The process whereby water vapour converts to a solid when cooled, omitting the gas stage. This can occur in reverse from solid to gas.

Supersaturation   When a pocket of air has a relative **humidity** of over 100 per cent, with enough water vapour to produce **condensation** but this has not occurred. For condensation to take place **hygroscopic particles** or **condensation nuclei** are needed.

Synoptic chart   The technical name for a weather chart which shows the **weather** over an area for a particular time.

Temperature inversion   An increase in temperature with height in the lower **troposphere** which is the reverse of the norm.

Thermal   A current of warm rising air, relative to the surrounding air.

Thermal inertia   The response of a body to a change in temperature. Land has a low thermal inertia, while sea a high thermal inertia.

Thermosphere   The layer of the atmosphere lying above the **mesosphere**. See **ionosphere**.

Transpiration   The loss of moisture from plants through pores (stomata) on leaves.

Tropopause   The boundary between the **troposphere** and the **stratosphere**.

Troposphere   The lowest layer in the atmosphere; it is about 9km thick above the equator and 6km thick above the poles.

Trough   An elongated area of low **pressure**.

Urban microclimate   See **microclimate**.

Vapour pressure   The partial pressure of water vapour, along with other gases in the atmosphere.

Vorticity   The amount of spin possessed by a rotating body – including air in the atmosphere.

Water balance   A similar idea to the water budget but applied at a smaller scale.

Water budget   A spatial balance at the global or continental scale between areas that experience net **evaporation** outputs and areas that experience net **precipitation** inputs.

Water cycle   See **hydrological cycle**.

Weather   The state of the atmosphere over an area for a short period of time. The **elements** of the atmosphere, e.g. **precipitation**, **pressure** and wind, are used to describe the weather. Both the temporal and spatial scales involved are smaller than when referring to **climate**, hence the use of weather forecasts and weather charts.

Index

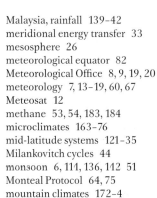

Acknowledgements

Published by Collins Educational
77–85 Fulham Palace Road
London W6 8JB

An imprint of HarperCollins*Publishers*

© 1995 Paul Warburton

First published 1995

Reprinted 1996

ISBN 0 00326685 0

Edited by Anne Montefiore
Designed by Jacky Wedgwood
Picture research by Caroline Thompson
Artwork by Barking Dog Art, Contour Publishing, Hardlines, Jerry Fowler, Malcolm Porter
Index by Laurence Errington

Typeset by Harper Phototypesetters Limited, Northampton, England

Printed and bound in Hong Kong.

Dedications

The heavens declare the glory of God; the skies proclaim the work of His hands
Psalm 19:1

Writing any academic book of this size involves obtaining a huge amount of information and data from many sources. I would like to thank all those who provided the excellent material on which this book is based. I am also indebted to my wife Eileen and my family for their support, patience and encouragement over the last year.

Much of the inspiration for this book comes from teaching my past and present A-level students and from conversations with other Geography teachers. I wish to acknowledge their inspiration, views and interest.

Finally, the excellent layout and presentation is the result of the efforts of Harper Collins staff. I particularly wish to thank Ela Ginalska (Publishing Manager) and Anne Montefiore (Project Editor).

Paul Warburton

Every effort has been made to contact the holders of copyright material, but if any have been inadvertently overlooked the publisher will be pleased to make the necessary arrangements at the first opportunity.

Photographs
The publishers thank the following for permission to reproduce photographs:
Bryan and Cherry Alexander, Figs 2.18, 2.19;
The Associated Press Ltd, Figs 8.8, 9.3;
D. Vaughan/BAS. Courtesy of EOSAT Ltd, Fig. 4.18;
BP Solar Ltd, Fig. 2.8;
Barnaby's Picture Library, Fig. 6.27;
Doug Scott/Chris Bonington Picture Library, Fig. 2.10;
J. Allan Cash Photolibrary, Figs 6.9, 7.3, 10.6, 12.18;
Colorific!/Telegraph Colour Library, Fig. 13.3;
Gerry Daly, Fig. 1.24;
Empics Ltd, Fig. 9.24;
Sue Ford/EPL, Fig. 5.14;
David Cumming/Eye Ubiquitous, Fig. 4.2;
Crown copyright. Reproduced with the permission of the Controller of HMSO, Figs 1.7, 1.14, 1.25;
Rob Cousins/RHPL, Fig. 13.9;
Tony Waltham/RHPL, Figs. 13.10, 13.12;
Holt Studios International, Fig. 1.3;
Hulton Deutsch Collection Ltd, Fig. 9.17;
The Hutchison Library, Fig. 6.1;
The Illustrated London News Picture Library, Fig. 3.7;
Andrew Lambert, Fig. 2.25;
London Aerial Photo Library, Fig. 9.18;
The Met Office/BBC, Fig. 5.24;
The Met Office/ETH Zurich, Fig. 8.17;
John Shaw/NHPA, Figs. 2.21, 2.22;
NOAA/NCDC, Fig. 11.13;
Neil Cooper/Panos Pictures, Fig. 7.14;
Sean Prague/Panos Pictures, Fig. 8.1;
Marcus Rose/Panos Pictures, Fig. 10.1;
Range/Bettman/UPI, Figs. 11.7, 11.8;
Range/Reuter/Bettman, Fig. 11.1;
Rex Features/Today, Fig. 9.20;
Science Photo Library, Figs. 1.8, 1.10, 1.11, 1.12, 2.3, 2.26, 3.1, 3.10, 3.12, 3.13, 3.14, 3.19, 5.1, 5.6, 5.8, 5.21, 6.14, 6.28, 7.1, 8.2, 8.19, 11.3, 11.12;
South American Pictures, Fig. 12.13;
J Berry/Liaison/FSP, Fig. 11.14;
Mark Edwards/Still Pictures, Figs 5.22, 10.12, 13.15;
Tony Stone Images, Figs 1.5, 2.1, 3.21, 7.18, 8.4, 8.5, 10.5, 12.1, 12.27;
Telegraph Colour Library, Fig. 12.15;
Bob Thomas Sports Photography, Fig. 1.1;
University of Dundee, Figs 9.14, 9.25;
Paul Warburton, Figs 1.5, 3.2, 3.8, 8.10, 8.12, 8.30, 12.3;
Zefa Pictures, Fig. 10.3.

Cover picture
Blizzard, Times Square, New York
Source: Tony Stone Images

Maps
Fig. 12.11 – Swiss Atlas, plate 65, edition 1973, editor, Erich Schwabe.
Reproduced by permission of the Swiss Federal Office of Topography, 9 March 1995.
Figures 10.21 and 10.22 are Crown copyright, reproduced with the permission of the Controller of HMSO.

Additional contribution
The author wishes to acknowledge assistance given by:
BP Solar
Bureau of Meterology, Australia
Martin Airey, University of East Anglia, Climate Research Unit
Manchester Weather Centre
Mrs D Holt, Librarian, Manchester Metropolitan University
NOAA, US Dept of Commerce
University of London, Institute of Commonwealth Studies
University of London, Institute of Latin American Studies
University of Salford
World Meteorological Organisation